Much Master T
One VJ's Journey
by tony young and dalton higgins

I dedicate this book to my wife Paula, son Kalif, mom Angela, brother Basil, and to all those who have gone before me, but still sustain me: my dad Karl Young, my grandmothers Maudrian Dawson and Myrtle Richards, my uncle Rupert Prince, my uncle Lloyd Campbell, and my cat Borris.

Copyright © Tony Young and Dalton Higgins, 2002

Published by ECW PRESS
2120 Queen Street East, Suite 200, Toronto, Ontario, Canada M4E 1E2

All rights reserved. No part of this publication may be reproduced, stored in a retrieval system, or transmitted in any form by any process — electronic, mechanical, photocopying, recording, or otherwise — without the prior written permission of the copyright owners and ECW PRESS.

NATIONAL LIBRARY OF CANADA CATALOGUING IN PUBLICATION DATA

Young, Tony, 1961–
Much Master T: one VJ's journey / Tony Young and Dalton Higgins.

ISBN 1-55022-541-3

1. Young, Tony, 1961–   2. Video jockeys — Canada — Biography. I. Higgins, Dalton. II. Title.

ML429.Y775A3 2002   791.45′028′092   C2002-902177-4

Cover design: James Oliver Senior
Text and photo section design: Tania Craan
Typesetting: Gail Nina
Production: Mary Bowness
Front cover photo: Steven Carty
Tupac Shakur transcript reprinted with permission of MuchMusic
Printer: Transcontinental

This book is set in Garamond and Franklin Gothic

The publication of *Much Master T* has been generously supported by the Canada Council, the Ontario Arts Council, and the Government of Canada through the Book Publishing Industry Development Program. **Canada**

DISTRIBUTION
CANADA: Jaguar Book Group, 100 Armstrong Avenue, Georgetown, ON L7G 5S4

PRINTED AND BOUND IN CANADA

ECW PRESS
ecwpress.com

Contents

1) The Early Years

From Leeds to Loonies   1
Master T (T is for Thespian)   19
Cleft Palate/Harelip   28

2) Life at Much

On-Air ID Phenom   37
MuchMusic Groove   38
X-Tendamix   49
VJ 101: How To Be A VJ Forever (Without Getting Tired)   57
Alter Egos   60
Master T for Premier: PC Convention Coup   68
Da *Dance Mix* Years   76
Da Mix: Da Early Years   83
The Real Deal   86
Soul Train Music Awards   91
T's VJ Primer   94

3) Racial Diversity and Urban Music in Canada

Invisible Minorities in Idiot Boxes 103

Happy to Be a Nappy VJ 110

Iced Out 115

Urban Music in Canada 117

Roots, Rock, Reggae 134

4) Celebrity Interviews

Celeb Hopping 145

Goodbye Blocko 246

Life After Much 256

Acknowledgements

First and foremost I give praise and thanks to the creator, for making all things possible. My career at Much has been an incredible journey and to have it documented at this point in my life is truly divine.

I'd like to take this opportunity to thank my foundation, who have stood by me throughout the years. A special thanks to my wife Paula: not only did you conceive the idea for the book, but your contribution has also been invaluable. I thank you for your tireless writing, insight, and attention to detail. Thank you Kalif for being so patient and understanding during this process — Daddy's so proud of you. Thank you Mom for being my biggest fan and for taping all my shows, which definitely aided us in the process of recounting the interviews. Thank you Dad for your spiritual guidance and for teaching me to always hold my head high. Thank you Basil for your support and professionalism on some of the biggest interviews in our careers. Thank you Aunt Sis for being my second biggest fan and for being in my corner. Thank you to the Johnson family (Donna Johnson-Huggins, Kirk Johnson, Allison Cooper, and especially Joy Reynolds) for all your love and guidance throughout my personal endeavours. Thank you to all my in-laws, nieces, and nephews (Kurt Huggins, Laurie Johnson, Karl Cooper, Yvette Sutherland, Travis Johnson, Tarin Huggins, Malayka Cooper, Nathan Johnson, Natalia Cooper, Omari Young) for inspiring me and making my life richer. Thank you Dave Campbell for showing me what friendship is all about and for your dedication to the music. Thank you Danny G for extending your friendship all the way from the West Coast, especially through the *Dance Mix* years. Thank you Dawn Craig

Much Master T

for seeing us through all the ups and downs; your kindness will never be forgotten. A big up to the old school crew (Fitzroy, Carlton, Mike, Brian, Everton, Byron, Hartnell, Chad, Ken, and Lloyd).

A special thank you to Jack David for believing in the concept from the very beginning and for this amazing opportunity. Another special thank-you to Jen Hale for working so closely with us through this gruelling but rewarding process; the book wouldn't be what it is without your guidance. Thank you so very much, Dalton Higgins, for pursuing the deal and for the solid job you did on this project, especially towards the political aspects of the book: a wicked first effort, brethren! Thank you Micah Locilento for your editing expertise and for articulately streamlining the flow of the book. Thank you Joy Gugeler for the promotional push and for helping us to devise a unique strategy for getting the word out. A big up and thanks to the ECW staff for all their help in making the book happen.

A special thanks goes out to Moses Znaimer for believing in me before I was ready to believe in myself. To all my friends and co-workers at the Nation's Music Station, thank you for the 17 positive years — without you all, I wouldn't have been able to realize so many of my dreams. A big up to the Canadian Music Industry and thanks for the memories while I was at Much. And last but not least, to the "Nation," "Beautiful People," "Cool Cats," and "Soul Mates": I continue to be blessed with your positive vibes, and I thank you all for religiously tuning in over the years and I hope you enjoy the book.

Foreword by Shaggy

Master T was a positive influence for all kinds of music.

I met Master T the first time I was ever in Canada, in 1993. I trusted him the moment I met him. To me, he very quickly became the face of MuchMusic. A trip to Canada invariably meant a trip to see Master T. He loved music and because of his vast knowledge of all kinds of music, especially urban influenced and reggae, it was easy for him to connect with all types of artists. He understood the journey and the music. He not only spoke our language but he was very connected to the streets. He was the first representative of a "mainstream" media outlet to introduce urban music and style to the masses. He was also very much in touch with reggae music and responsible for the popularization of both urban and reggae music in Canada. I was born and raised in Kingston, Jamaica and reggae music was in my blood. Master T understood this music and the culture that surrounded it and helped bring this culture and music to the Much audience. I not only value our relationship with respect to the music, but I also value Master T as a friend.

MuchMusic helped expand the reach of urban music in Canada. For the first time kids had another means to this type music. At the time, Much was already an important vehicle for mainstream/top 40 music, but it was Master T's influence that first gave urban music a voice and a presence. He was the chosen representative for the urban music community and it quickly became clear to the Much audience and to the artists that if something was happening, Master T would be the first to know and in turn share that knowledge with the Much audience. He brought the "streets" to the masses. Because of his vast knowledge of all types of music, artists trusted him and valued his opinion. He took the time to understand where individual artists were coming from and how they fit into this new world called music television.

This relationship I developed with Master T was very special. I looked forward to each visit to Much; in fact, I would all but refuse to be interviewed by anyone else. I have been doing this over 10 years and Master T was there from almost the start.

I thank Master T for his work and his friendship.

— *Shaggy, October 2002*

The Early Years

FROM LEEDS TO LOONIES

When you've been in the public eye for 17 years running, folks tend to think that they've got the private side of you all figured out. But I'd say that perception and reality are two radically different things. Born in Leeds, England, in 1961, I grew up in the Mexborough area with my older brother Basil (who's two years my senior), my mom Angie, and my dad, Karl. You know how they say somebody's personality is carved out by the time they're five years old? I'm living proof of that. Straight up, I was born to entertain.

My parents were both born in Jamaica and left the island by boat in their early 20s in search of a "better life," as did so many other Jamaicans at the time. In 1955, my father was the first to make the journey and my mother followed two years later. I recall my mom

Much Master T

saying that she had a thing for my dad when they were both teenagers, so when my father decided to go off to England as a young man, it wasn't so much that she wanted to leave the Caribbean and move to Leeds to follow her dreams. In actuality, she was more interested in following her dream man.

The only obvious downside to my childhood was the overt racism my family and I experienced in England. If we weren't surrounded by so many other transplanted Jamaican families who acted as a support unit and safety net for each other, who knows what would have happened. I'd say it's because of my close connection to this Jamaican expatriate community in Leeds that I'm still so in touch with my roots and culture today and can speak the Caribbean dialect (patois) fluently. My parents also stayed true to their original Caribbean tastes and — aside from Canters Fish and Chips (the number one local fish and chip shop in our neighbourhood) — the main staples of my diet consisted of yam and banana, fried fish, curry chicken, and a few other things Caribbean.

Although I grew up to be a solid six feet, two inches, I was a very tiny, fragile tot. A lot of folks that my parents used to hang out with in England have told me that they used to carry me around in a basket 'cause I was so small. But that didn't stop me from raising a ruckus whenever possible. As a toddler I would get into all sorts of trouble and my mom likes to tell stories of how I'd give a lot of "back chat" (this means to "mouth off" in patois) to whoever ticked me off. One particular incident that sticks out in my mind occurred when I was only three years old. It was Christmas time and our family had taken a trip out to see Santa at a local shopping centre. You know the routine. You wait in line nervously until it's your turn to sit on Santa's knee. My big brother Basil got the call before I did.

The Early Years

So he sat on Santa's knee and told this big, furry white guy what he wanted for Christmas. Santa complied, as most Santas do, and he reached into his big bag of toys to pull out something that he gave to Basil; my brother accepted it graciously. When my turn came around, I went up to Santa and jumped up on his knee and asked him for a fire engine. Santa, being the North Pole scholar that he is, dug deep into his bag and pulled out a gift. I grabbed it out of his hands and opened it immediately. I was disappointed to find out that it was not a fire engine, but a stick and wheel. Devastated, I lugged this meaningless gift around the shopping centre for hours. As soon as I got outside the mall, my mom tells me, I yelled, "stupid Santa, this isn't a fire engine!" and hurled the toy clear across the other side of the street without looking back.

Over the years, I've come to realize that this take-no-guff-from-anyone (even jolly old St. Nick) personality trait spans three generations. I inherited it from my father, and my son has an identical disposition. When I wasn't busy questioning the relevance of St. Nick, I was a very caring, loving, and emotional child that people were naturally drawn to. I guess you could say that I was a textbook mama's boy who was unapologetically overprotective of my mom, Angie. That's not to say that I didn't have an extraordinary connection with my father. Because I loved them both independently of each other, I was better able to handle the news that my parents were getting a divorce when I was only nine years old. Sure, I felt a lot of sadness and an overwhelming amount of anger when my parents divorced, but I didn't implode right away because, fortunately, I had a few escape mechanisms built in. I'd play a lot of soccer or just act out with my friends to ease the pain. My family says I have an "old head," which simply means that I had a certain maturity and

Much Master T

wisdom about me as a child. It was almost as if I had been here on the planet before. That doesn't mean that my old head made the impact of my parents' split any easier. The fragments splintered off into many aspects of my development as a teenager.

When my father relocated to New York in 1968, he continued to work in the electronics profession. He would write to us and continuously stay in touch. After my parents split, I gained an enormous amount of respect for my mom. I have always been attracted to strong women — probably because I had so many strong women around me while I was growing up. Myrtle, my grandmother on my mother's side, had an innate wisdom and a peaceful, vibrant strength, while my father's mother, Maudrian, had a powerful aura that was paired with a certain determination. My mother, who became a nurse in 1970 and single-handedly raised my brother and me after my father moved to the United States, continued to be a pillar of strength through all our ups and downs.

We were a tight family unit. I remember my mother taking us on short, inexpensive vacations to the beach in a caravan (England's version of a mobile home); we had a lot of fun just doing the simple things in life. One year she sent us on a memorable trip to visit our father in the Bronx, New York, during our summer holiday. The trip was designed to get us caught up with our father and re-establish the bond we were missing so much. One thing I'm thankful for when it comes to my mother is that she never tried to poison our minds and turn us against our dad after they went their separate ways. My father was still able to be a part of my life, just in a different way, via long distance. He was just a phone call away and I carried his energy around with me all the time. My big brother Basil became the "man of the house" in Dad's absence and he took this responsibility very

The Early Years

seriously. As a kid it was difficult to shake loose of my brother's overprotective influence on everything I did, but as we became men, I began to understand what a positive role he has played in my life. It was during this time in the early '70s that my mom began the arduous process of applying to the immigration authorities to have us sponsored by her sister (my Aunt Sislyn) and brother-in-law (Uncle Rupert) so we could move abroad.

One of the most important turning points in my life was coming to Canada from England when I was 13. When our family of three emigrated from England to Kitchener-Waterloo in 1974, I wasn't the least bit prepared for what Canada had to offer. Relocating from a small town like Leeds to a city like Kitchener, where everything is so spread out, was difficult. As a child, the upsetting reality of having to leave my best friends back home in England — Kevin, Tony, Michael, and Troy — was just as traumatic as finding myself in a new country. I enjoyed playing soccer with these guys, and in our young, underdeveloped minds we all thought that one day we'd play together at the pro level — on the same team, of course. Sure, it seemed ridiculously infantile of me to dream this, but when I tried out for Leeds United's junior apprenticeship squad, I felt I had a legitimate shot at making the team. To a kid with dreams of football stardom, England seemed like the land of milk and honey, if only because they produced good soccer clubs. But in retrospect, if I had stayed in England I doubt that I would have been able to pursue a broadcasting career, due to the way their elitist and classist educational system is set up. The system has a terrible influence on the living standards of English people of colour.

There was definitely a large degree of culture shock when I physically arrived in Kitchener. I remember the drive from the airport, and seeing all of these big, schmancy cars. I was always fascinated by cars, and seeing all of these oversized cars represented what North America was all about: pomp and pageantry. Everybody had phones, swimming pools. It all seemed so grandiose.

Because my aunt Sis and uncle Rupert had sponsored us to come to Canada, we ended up living with them for over a year before getting settled in our own home. Uncle Rupy was one of my mentors and he was like a father figure to me. He offered up amazing pieces of advice, tidbits of info, and shared life lessons that I took to heart, which still apply to this today. I'm of the opinion that our elders don't get the credit they deserve for all of the sacrifices they endure in order to provide us with a better life. All the adults that surrounded me as a child showed me, by example, an amazing work ethic that made a lifelong impression on me.

My first day at Queensmount Public School in Kitchener stands out in my diminishing LTM (long-term memory). It was the first time I had to confront my culture shock. That chilly, overcast afternoon, I remember walking around in the schoolyard to get a feel for my new environs and my new school. All the other kids were standing outside and I was scared stiff because I had only made one friend since arriving in Canada. His name was Brent Stamp, and he lived right across the street from us. Unfortunately, he didn't go to Queensmount, so I had to go it alone.

I was terrified by my first real interaction with so many Canadian kids my age. I saw a couple of fellow black kids on my first day, and I remember glancing over to one of them, thinking to myself that I have to go for the common element here. Because the two of us were from the same race, I figured this guy would be a good bet to show me the ropes. Feeling like an alien with a funny

The Early Years

accent, I badly needed some companionship. So I walked over to him with my thick Brit accent and said, "My name's Anthony Young. I'm from England." The kid just gave me one of those looks that said "this guy can't be serious" and then he told me to "get lost, man." Boy, was I devastated.

Back then, the public school fashionistas were also out in full force. As I was walking into school my first day, I have a vivid memory of my schoolmates taunting me by chanting "floods, floods." Of course, I had no idea what they were talking about because I had never heard that term in Leeds. I figured maybe they thought that, because I was from England where it seems to rain every day, I was used to seeing the streets flooded with water. But the real story is that my mother had bought me these grey dress pants to wear to my new school and at the beginning of summer they were just fine but I had grown three to four inches over the summer and now they were up around my ankles. It took me a while to figure out what this "floods" business was back then, but it didn't take me any time at all to catch on to the latest trends from a Canadian junior high fashion perspective.

When I arrived at Queensmount I was placed in the seventh grade, which meant that I had actually been held back a year. In fact, when the school administrators looked at my report card and saw that my grades from England weren't of the Rhodes Scholar variety, they strongly suggested that I attend a vocational school. This was alleged to be the standard practice for youth of colour with lower-than-average grades who immigrated to Kitchener-Waterloo in the '70s. Thankfully, my mother fought this oppressive race-driven streaming process and I was allowed to remain in a regular public school.

Much Master T

After finally getting to attend Queensmount Public School — only after my mom insisted — there were other silent protests going on as well. I was waging my own personal war against the system because deep down I was still missing England. I refused to stand up for the Canadian national anthem, which is a major faux pas in the public school system. The only Canadian custom I did engage in was eating hamburgers, French fries, and macaroni and cheese. As for all the other Canadianisms, I wasn't interested because I simply didn't want to be here.

At Queensmount I had this teacher named Mr. Armstrong who really took a liking to me. I sensed that he saw some potential in me and he helped to nurture and develop my love of theatre. He insisted that I get involved in the theatre program at school. So how did I repay him? By being even more of a shit disturber in his class. Because I knew he favoured me as a student, I felt I could get away with anything. I could sense that he was really frustrated with me but he didn't let on as if he was. When our school staged a rendition of *Welcome Back, Kotter* (I was in the eighth grade at the time), Mr. Armstrong recruited me for one of the leading roles. Can you guess who I played? Washington, of course. He was the tall, cool, jive talkin' black dude. Naturally, playing Washington for a week raised my profile at school and made me more popular among my classmates. That's when the acting bug bit me for the first time.

You're probably wondering how I landed this socially lucrative Washington gig, huh? Well, there was this classmate of mine named Jay who desperately wanted to be elected as either the president of the school or of the student council. As part of his campaign speech, he enlisted my services as a political spokesperson of sorts. Right after his public address, I went up on stage and did this imitation of

The Early Years

J.J. (Jimmie Walker) from a popular '70s sitcom called *Good Times*. J.J. was similar to Washington from *Welcome Back Kotter*, but he was more verbose, geeky, and obnoxious. So I went up on stage and said, "J.J., he's alright, but for co-president he's gonna be Kid Dynomiiiite." The audience thought my J.J. schtick was hilarious, but deep down I knew I was just acting the fool to fit in. So it was this pathetic performance that apparently led Mr. Armstrong to recruit me for the part in *Welcome Back, Kotter*. Participating in theatre helped me focus on the social networking aspect of public school life in Canada. Nobody wants to be an outcast or a misfit, and this was my first taste of the spotlight in my new school. It felt good to be considered a capable Washington.

As far as my Brit accent was concerned, all of my peers enjoyed my accent so much that they insisted I recite everything in English class. Even the teachers themselves loved my accent; it was exotic to them. But I didn't want to be the exotic Brit. I wanted to fit in — not stand out like a circus clown — so I started to Canadianize myself. Diving headfirst into the world of Kodiak boots, butter tarts, Levi Strauss, lumberjack shirts, McDonalds' hamburgers, and all that living in North America had to offer, I tried to conform really quickly. As it turned out, my accent was gone by the end of the year.

Heading into my high school years, I was more confident in my abilities (both as a "Canadian" and otherwise). I attended Kitchener-Waterloo Collegiate Institute (K-CI), and if you're expecting to hear some uplifting story about how I got straight A's and always knew I was gonna be a successful VJ, you might want to skip over to the next chapter. I didn't care much for schoolwork. Playing sports, being cool, and picking up hot "babes" was what I was there for; homework was definitely an afterthought. I was a straight B-/C+

Much Master T

student who scored consistent A's in congeniality. K-CI was a decent school that had a semester system in place, which meant that students could gear their course loads toward one specific area of interest if they wanted. For example, if you knew you wanted to be a doctor from the first moment you stepped foot in K-CI's hallowed halls, you could immediately put yourself on a career path.

In grade nine, my favourite subjects were math and science, so one day, after a little pondering, I decided that I wanted to be an electrician. That interest faded fast, and by grade 10 my shiftless adolescent brain had me convinced that I should become a performing artist, so I geared all of my courses towards the arts. My own carefully crafted curriculum consisted of math, dominoes, gym, and uhhh . . . more dominoes. By the time I hit the eleventh grade, I had become one of those slacker teens you read or hear about, skipping class all the time to hang out in the cafeteria. Let's just say that my report card was less than stellar. Of my seven credit courses, I managed to pass only two. This meant I had to attend summer school out of necessity, not choice. But as fate would have it, I had taken extra credits in grade nine, so getting caught up academically wasn't a complete mess. Strangely enough, the two credits I did manage to pass in grade 11 were both related to media and mass communications.

From early on, I seemed to take a real interest in broadcasting, though I wasn't really aware of it at the time. I believe that everybody has a gift for something — by nature or nurture — and it's just a matter of tapping into that gift and pursuing it. During my last year of high school I had started volunteering at the local cable outlet. This broadcasting thing must have been in my bloodlines; my brother Basil was studying Broadcasting at Niagara College, and

The Early Years

my uncle Lloyd Campbell, who lived in the States, worked for NBC. He was an Emmy Award winning editor who had been recognized for his work on *The Young and the Restless*. In fact, whenever we'd visit Uncle Lloyd he'd give us tours of the large NBC facilities. I still have pictures of us as teens hanging out at the NBC studios in New York. I guess the experience stuck with Basil and me, as we both went on to have long careers in TV land.

It became clear to me that if I didn't really enjoy something, it would be difficult for me to apply myself. School, at the time, was never really my thing. I can also thank the disco music trend at the time for tossing my inner Ivy League aspirations out of the window. When I was the same age as many of my Much fans, I would sneak into nightclubs illegally. Although you had to be at least 18 to get through the door, some of my friends (like Carlton, Fitzroy, Mike, and Brian) and I would somehow pass for legal age and be allowed in. There was one club in particular that we all loved, and it was called Fat Albert's. Ivan, the resident DJ, was absolutely amazing. He'd spin hit tunes from disco legends such as Gloria Gaynor and Cheryl Lynn and blew us away with his incredible mixing skills. So we'd go clubbing religiously every Friday and Saturday just to hear him spin.

Another thing about going out to clubs is that you had to dress to the nines to stand out. I spent way too much of my adolescence looking for nice outfits to wear out to clubs. I used to buy copies of GQ magazine and mimic the outfits the models wore. Basically, I'd see a Giorgio Armani suit that would be worth $1000 and I'd try to achieve the same look (minus a few zeros of course). I'd spend my valuable time shopping for nice, slick, pointy-toed shoes at a specialty store on King Street in downtown Kitchener called Salamander. Despite the fact that I was a teen, I was big on the sartorial elements

Much Master T

and I even wore a pocket watch so I could get that fine übergentleman look. My father was definitely a clotheshorse and he revered quality and cut in his suits. He passed that trait on to me and my brother as children. He would always say, "Only the best for me and my family; you must take pride in your presentation." And it stuck. Back in the old school days this was true of a lot of Caribbean folks; we've always been into creating our own unique look and it was far more important to us to develop our own individual style.

Something had to support my foray into the world of high fashion, and that meant that I always had to have a job in high school. Obviously, it was important to rake in some dough in order to stay fashionable, and I always felt responsible for making a contribution to the household. One year I landed this job at the neighbourhood gas station. Another year, I lucked out in landing a gig at the hospital where my mother worked. I was part of the housekeeping department and it was a good experience that taught me responsibility, discipline, and punctuality, but most of all I was making big bucks: $9 an hour in 1979.

I didn't spend every second of my high school years working. During my down time, I was able to socialize with the "brothers" (other young black males) who had shared similar life histories. When we weren't busy thinking of ourselves as the hottest thing since sliced bread, we'd be hanging around the radiator in the basement of K-CI listening to burgeoning hip hop acts like Sugarhill Gang and Grandmaster Flash on our ghetto blasters. It meant a lot to me to be around fellow blacks because since I had left school in England (where the student population was 80 per cent black), I had always gone to schools with considerably few black students. I needed the comfort of being around people who had similar lifestyles, habits, and interests.

The Early Years

It was during these late high school years that I would become more acquainted with my eventual wife, Paula. Actually, we first met at a summer basement party soon after I arrived in Kitchener. For a number of years we were just casual friends. A particular group of girls used to consistently come out to our soccer games at K-CI and Paula would hang around with them. When you couple my love of soccer with my Brit roots, playing pro football was not entirely out of the question. So I funneled that energy into displaying my skills on the K-CI senior boys soccer team. I played one of the forward positions on the varsity team and we won the city championships one year and lost in OFSSAA (Ontario Federation Secondary Schools Athletic Association) another year. It was during this time I heard through the grapevine that Paula had a crush on me. I was one grade higher than she was, but one year we somehow ended up in the same math class together. She sat across from me and I would talk to her about my family and my grandmother Myrtle, who had just come to Canada from Jamaica. It was funny; from the first time I met Paula we had so many similar interests. She, too, had come from a divorced family and her three siblings had been raised by her mother. Like me, Paula was considering a career in acting during high school, and she had done some performing arts in public school before that. We both performed with The Caribbean Association in Kitchener. It wasn't until later on in the game that we realized we were soul mates.

Of course, having a potential soul mate in one of my classes didn't mean that I was present for every session. Our math teacher had an "open door" attendance policy that made it easy for me to go to class without actually having to sit through any lessons. "If you want to come here and leave at your own discretion," he would tell the class, "it's entirely up to you." Of course his words weren't

Much Master T

intended to be taken literally, but you couldn't tell me that. I'd check in when the class began and then leave the classroom to walk around the halls, chat with a few people on their spares, and then return to the class at my own leisure to learn the day's lesson. Once a week we would have a surprise quiz and I would finish in about 10 minutes flat. Antics like this confused Paula because she couldn't figure out whether I was a genius with a high IQ (who could ace exams within minutes) or just a total cop-out. Well, it became clear that I was definitely more the latter. To nobody's surprise, this was one of the classes I failed. Despite my high school shenanigans, I did manage to graduate on time, in the summer of 1980. In the course of my "studies," I had mutated into this social butterfly who didn't mind being the life of the party. I am eternally grateful to Mr. Armstrong for being such a positive force in my life, for introducing me to extra-curricular pursuits. Years later, when I was on MuchMusic, I returned to his school while he was teaching a grade six class and talked to the kids about literacy.

Although high school in Kitchener was a blast, it was time for me to go to college. I was pretty linear in my career scope; I wanted to study television and nothing else. I wasn't even interested in radio at the time. I had heard that Mohawk College in Hamilton offered a pretty concentrated course of study in this area (as did Seneca and Fanshawe), so I applied to all three. It was a frightening period for me because it's hard to prepare oneself for any other options but going to college. What would have happened to me if I hadn't been accepted to college? In the '80s (and I'd assume now as well) most Caribbean parents insisted that if you weren't going to university you had better find something just as fruitful to do with your time. Wisely, my mom determined that I had better take that one-way

The Early Years

ticket to Somewhere University or Anywhere Community College. As she figured, they'd come to a new country and sacrificed a lot so that my brother and I would be able to take advantage of the educational opportunities that she and my dad never had. In fact, after I failed all of those credits in high school and, at one point, threatened to take a year off to work, my mother wasn't initially interested in me settling for a mere community college, but in the end she was cool with it.

When I completed the interview to get into Mohawk's broadcasting program it was a big accomplishment for me in the micro sense. It gave me a feeling of independence. My brother was already blazing family trails studying cinematography at Humber College. And despite popular perception, it wasn't easy having a cool, confident older brother. If I thought I was popular at this time with the fellas and the ladies, he was The Man. Once I was finally accepted into the broadcasting program at Mohawk, I boarded with a family for my first full year of college. I wasn't completely comfortable with the idea of leaving home and being completely dependent on myself. I couldn't cook that well, I was lousy at doing things like laundry, and who was going to ensure I wasn't going to party myself into oblivion? I lucked out because a black family took me in and the eldest daughter Gail was close to my age and we became fast friends. Her mother had high standards and taught me so much about perseverance. She was a nurse by trade and was also attending university at the same time. Gail's little brother Alan had a bratty streak in him that reminded me of my own behaviour as a kid, so we understood each other.

A month before the end of my first year at Mohawk I went back home on the weekend as per usual. But this time I decided to go out

Much Master T

and entertain myself for a change. It just so happened that I ran into Paula at a party in Waterloo. Talking and dancing together for most of the night, I could tell that she had been through a lot of changes. And I guess I had changed as well. Somehow, I thought my first year of college was gonna be all about getting laid and partying all night, but that just wasn't the case. I started believing I was socially inept and that something must be wrong with me because women weren't digging me. There was no wild sex going on between two mutually consenting college-aged adults. Nada. Zippo. Zilch. Even my less attractive friends (in my mind, anyway) were getting their National Lampoon–like freak on. I became introspective overnight and I started imagining what qualities I really needed in a potential mate. I had a lot of time to do this because it's not like my phone was ringing off the hook. I used to sit up late at night and make mental lists of the qualities and attributes that I wanted in a girlfriend. And the more I combed through these lists, the more I thought about Paula. I had just come out of a relationship prior to the start of my first semester in college, and that relationship didn't work out very well. So already, I was starting to give some more thought to the uncanny connection between Paula and me. Since I had gone away to college and she was still completing her last year of high school, her crush on me had somewhat fizzled and she had moved on to other pursuits. She also felt that if I was really interested in her at all, I would have made a move ages ago. So now I had to work my butt off if I was going to get her to take me seriously as a contender, much less go out with me.

My first year at college was very clique-ish. You'd have this one core group of cool dudes, the "in crowd" that everybody admired and wanted to hang out with. I ended up hanging out with this guy

The Early Years

named Johnny who was a massive punk rocker. He'd have pins in his ears, in any and all appendages, and in his pants. He was a total radical. I had my hair braided, so I didn't fit in with the mainstream of Mohawk College life either; we were a good fit. As far as the racial kinship argument goes, the black men in Hamilton were a little bit on the snobby side, and Johnny was more at my grunged-out speed. Besides, I was going through a little depressive mode with all the changes in my new life. But it was in college that I promised myself I'd make my own mark and do things my way, like Sinatra.

The way colleges weed out slackers is a little bit intimidating, to say the least. In first year, they whip out these prepared speeches about how most first-year classes of 35 get whittled down to 16 by the time graduation time comes around. But I was up for the challenge. As I said, because I wasn't getting any, ahem, major "action," I tried to focus some of my need-to-get-laid energy on my schoolwork. In my "Cinema and Social Change" class, I put together a very cool slide presentation on the life story of Louis Armstrong, a musician I was fond of. There was very little else to focus on for me, because the Mohawk College party scene was a bit lame. I'd walk into these fraternity parties and guys would be running around with lampshades on their heads, inebriated out of their minds. I'd bring my own six pack of beer and they'd say, "Here's T with his poverty pack." That was really annoying, so I hung out for a few of the parties, but soon lost interest.

By my second year, I realized that the Prophets of Doom working in the administrative department at Mohawk College were right. I knew that my friend Johnny wasn't going to return to school in the fall because of this weeding out process; it was very disappointing. Don't get me wrong, I had other interesting friends, like

Much Master T

Andrea and Judy, but my closest friend during the course of the second year was Willie — I like to call him Face. We used to chug a few at the Arnie (Mohawk's pub) after hours and we began to realize through talking that the educational infrastructure that grooms specific people to do specific things was definitely in place here. It was in school that the administrators determined, amongst themselves, which students are CHCH TV or CBC TV–worthy, and I could see that only certain students were being groomed for these positions. To make things worse, I really started to miss my mom during my sophomore year and, as usual, things were often financially tight for me — due to the delays in my OSAP loan monies coming in on time every fall. To pass time and earn some extra money, I landed a job working in the library. As it turned out, this was a good gig for me because I would call Paula all the time long distance on the phone for free. Now it took some skill but we finally began dating exclusively on April Fools' Day 1981.

My most memorable part-time job was as a prep cook at the Keg restaurant on weekend mornings. During these lean college years, on occasions when I was really starving, I would heat up a piece of the Keg's delightful apple pie and cover it with a couple scoops of their creamy vanilla ice cream. The scam here was to gorge somewhere the boss wouldn't be able to see you. My hideaway was the walk-in freezer in the prep area. I'll never forget the day my manager opened up the freezer door and caught me black-handed. He looked at me and asked why I was eating a dessert that wasn't even an item on my prep schedule and I looked at him with my big brown eyes and whispered, "I was hungry." From this day on, the manager seemed to clue in to the plight of the broke college student. I can

The Early Years

remember times after this when he would come in and cut everyone a handsome piece of pie, top it with ice cream, and we would all chow down.

MASTER T (T IS FOR THESPIAN)

The acting bug bit me way back in junior high. But it wasn't until I was studying at Mohawk College that it really sank its teeth in. Early in my second year, our college theatre company was auditioning people for a play they were putting on called "No Strings Attached." I remember that, as I was preparing to audition for the part, Paula was giving me tips on how to capture the lead role. I did eventually get the lead part, and with the support of June (the producer of the play), I worked my butt off to bring the character to life. I had to because there wasn't a whole lot of time to rehearse before the first scheduled performance. You see, the entire play is written around the lead and his job is to narrate and carry the audience through a number of little vignettes that occur throughout the play.

Opening night came quickly and I knew I was ready to put on a good show. Unfortunately, Paula was not able to make that Thursday's performance — or so I thought. It was a sold-out show, but (unbeknownst to me) she had taken the bus down to Hamilton a day early. As it turns out, she was in the crowd to take in the opening night performance. When it was all over and I came out of the theatre to see a bunch of my school buddies that had attended,

Much Master T

I was shocked and excited to see her there. As usual, she was honest with me about what she saw. She stressed that it was a good performance, but she told me that I never really had the audience in the palm of my hand. This was something I needed to work on if I were going carry them throughout the performance. Considering that my brother, my mother, her mother Joy, and her older sister Donna were all making the long distance trip to Mohawk to see the play the following night, her critical review was more than welcome at this point. The next night, I was mentally prepared, focused, and finally able to get into the acting zone. I really needed this performance to be good 'cause — even though you might think they're "just family" — I was performing for some of the biggest critics apart from Siskel and Ebert. That night, I put on the show of my life! It was Broadway as far as I was concerned. I really blew everyone in that theatre away and it was from that moment on that I knew I wanted to be involved in acting.

With my career sights now set on acting, my third year of college became critical. School administrators would groom students to work on the Andy Awards ceremony, which was our in-house college awards show. The stakes were higher now, and the student voted to be the producer of this show would usually land a gig at CHCH TV. I think my instructor, John Bradford, knew that my aspirations were more geared toward a looser style of television and he always encouraged me to go for what felt right. After a vote conducted by my broadcasting class, it was decided that either myself or my classmate Dwayne should produce the show. The vote was a tie, and I won the re-vote, but for some reason they gave the gig to Dwayne. Maybe they felt that he was more the CHCH TV type. I took it all in stride. As a consolation, I was given the job of station

The Early Years

manager for the college TV network. Man, I was moving on up in the world like the Jeffersons. I even bought a briefcase with my $200-per-semester honorarium and pulled a major coup, getting soap operas shown on our closed circuit TV throughout the school.

As part of the Mohawk program, students in the upper years had to do an internship to get some hands-on experience of what it's like working in TV land out in the real world. My first job placement was at Channel 47 CFMT in Toronto. I worked on a soccer show with this guy named Dale Barnes and the producer's name was Vack Varikitis. It was a great experience for me because I got to work at a mid-sized up-and-coming station and immerse myself in big city living. When I started out, CFMT had me doing jobs like answering phones and opening up mail, but even that felt good because I was part of a new station that was about to bloom. It was also great to be in a workplace where every ethnicity was represented.

When I finally graduated from Mohawk in 1983, I got a job at CFMT and it was a great accomplishment for a plethora of reasons. For one thing, only 14 of the 26 people who entered the TV broadcasting program went on to graduate. My first paid TV job was in master control, where I worked as a playback operator for a salary of $12,500. The funny thing is, I landed the CFMT gig two months before I officially graduated. I made it a point to voluntarily go back and work over my March break after placement and when a position became available they offered it to me. My instructors graciously waived the rest of the year for me, though I still had to write all my finals. It didn't pay much, but I was lucky to have gotten a paying gig in Toronto — as opposed to the far-off reaches of Anywheresville. Paula was studying fashion design at Ryerson in Toronto at the time, so this job was a godsend; it was becoming increasingly difficult to

Much Master T

handle the long-distance relationship. When my job at CFMT became more secure, we were finally able to look forward to being together more often. Despite the fact that Paula was living in an apartment with her roommates in downtown Toronto, she was always hanging around my new place on Seaton St. in the Cabbagetown neighbourhood of the city. By the end of her last year at Ryerson, she had officially moved in. With my hefty salary of $12,500 going to rent, car payments, and other miscellaneous costs, my budget was thong-tight. Plus, my Friday work shift left a lot to be desired. It started at 5:30 p.m. and ran right up until 5:30 a.m. the next morning.

While I was working at CFMT, I finally had head shots taken. I was hoping to line up an agent so I could score some acting work, and (believe it or not) one of the first things I landed was a role in *Police Academy 3*. Now here's the thing: I was the stunt double for the actor Michael Winslow, who was one of the main characters in the movie. With no experience as a stunt double, I got the job through a hook-up from a friend, Kim Saltarski, who was a year ahead of me at Mohawk College. Sometimes, I now realize, you have to stop and question yourself when you get caught up in these things — especially when you're not certain you'll be able to handle it. But I had made the commitment and there was no turning back. Although I had never been on a jet ski before, I was really convinced, at the time, that I could get by on my athletic ability alone. But these machines take practice and the powers that be didn't have all day for me to learn. Let's just say that my skinny little ass became well acquainted with Lake Ontario several times that day. I wiped out so bad one time that I drank up half the lake while they were trying to shoot a bunch of us in a V formation. All I could hear was "Cutttt"

The Early Years

booming across the water. Having swallowed up all the dredge one man could possibly handle, I got sick to my stomach. Let me tell you, that was an experience I'll never forget. It taught me a few things about saying "yes" to things your body knows you can't do. Still, they got what they wanted on the last take and when the paycheque came in (at $1,500 per day for two day's work), I was rich.

They asked me to return on the third day but I declined because I had already lined up another acting job. It was on a video shoot for Ian Thomas's "Harmony." John Candy, Eugene Levy, and Ian's brother, Dave Thomas, were also in the video; I was beside myself that day, just hanging out with these greats on the set. Even though I was missing out on another $1,500 cheque, I tell ya, just getting the opportunity to see the original SCTV cast do improv and ham it up right there in front of me was priceless. After those two parts came my way so early, I was really excited to go full force into the acting thing and felt pretty assured that even more gigs would be coming my way. Boy was I wrong. The phone didn't ring for months and I started to get really discouraged. It was also very frustrating back then because the majority of the roles that were offered up for black actors were stereotypical parts like pimps, drug dealers . . . you can fill in the blanks. I began to think that maybe I wasn't good-looking enough; maybe my teeth weren't straight enough. I was driving Paula nuts with all my complaining. She estimated that it would take about five years for me to achieve anything in the business, so I went to work on paying my dues. During those years, I got a part in the motion picture *The Believers*, appeared on *Street Legal* (a Canadian TV series), and I starred in a Jamaican comedy called *Boysie and Imojean*, written by Devon Haughton, a popular local dub poet. Everything I earned from that play went into buying

Much Master T

Paula's engagement ring. I also did a made-for-TV movie, though I can't remember the title (that's how memorable the movie was). I kept pretty busy during that time but my very first on-air gig was on a show right there at CFMT.

I met Daniel Caudeiron, one of the most influential urban music personalities in Canada while working at CFMT. He was the host of *Black World*, a great TV showcase for black music, art, and politics. I liked the show and I wanted to offer him my services. I bugged Daniel for months to see if there was any way I could help out. It almost got to the point where he wouldn't even return my calls. One day, I confronted him in the hallway and asked why he was trying to avoid me. Here I was, offering my services to his show for free, and he didn't want to take them. Well, after my little rant and some reasoning he wound up offering me a job as a reporter. I felt I had entered the big time now. A young filmmaker named Stephen Williams (who'd ultimately go on to study under Norman Jewison and direct *The David Milgaard Story*) used to do the movie reviews for the show at the time. In my first reporter's job I used to pick and choose the personalities I wanted to interview for the show. I loved conducting interviews with up-and-coming talents who were still flying slightly below the radar and whose star would eventually rise. Some of my first major interviews were conducted with Kevin Pugh, a ballet dancer from the National Ballet of Canada, and Atlee Mahorn, a local track star who ran the 200- and 400-metre races for UCLA and who would represent Canada in the '84 Olympic Games in Los Angeles. Although I was starting to get some notoriety for my urban culture reportage, I started to realize that, after a year and a half at CFMT, my earning potential was going nowhere fast.

The Early Years

Daniel Caudeiron, Walt Grealis Juno Award recipient for lifetime achievement

I knew him when he was called Tony Young. I was doing a show which I had started off working with in 1979 called *Black World*. We were doing the show weekly and while I provided access and contacts to participants in the actual on-camera, on-air program, the technical staff we used was provided by the multi-lingual TV station CFMT. Back in those days there were stations that had been assigned by the CRTC (like Channel 47) [that] gave a lot of opportunities to people of different ethnic backgrounds, both in front of and behind the camera. Tony and I became acquainted on a weekly basis when I would go down to tape the show that I was hosting. After a normal day's work for people on the show, sometimes Tony would be there handling one of the cameras, helping light the set, and we would josh and joke and banter. We'd sit there in the edit suite and exchange stories about the music scene, club scene, because I was doing all of these entertainment stories and I knew everybody through the Cheer DJ dance pool that I ran and still run.

°fi°fi°fi°fi°fi°fil'd have to get dressed, made-up, and come on set to do my interviews with subjects for the show and Tony would be on the camera taping that final one-hour show to be aired later on in the week. Gradually what happened over time is that he began to become more helpful as far as coming up with ideas. He'd be pitching me ideas about things happening in town and suggesting that we book cameras for certain events. So what you have is a young guy who had his ear to the street and the

Much Master T

club scene. He was not just doing CFMT's appointed set-ups and regular job duties, but he would go out with his own ideas in mind, on his own time, so it wasn't me saying to him, "Tony, cover the Minister of Tourism from Trinidad when he's in town." I got him to cover events for me because it was natural for him, he could work on his own pieces without my interference because he already had the editing skills to put together a little two- to three-and-a-half minute clip about any event. This is what he began to do and this is what [he was] a part and parcel [of] in '82 to '83. He was a part of this influential young crew of upstarts who went on to do big things. Stephen Williams, a guy who has become well known as a film-maker and cinematographer in Canada, he was a student at one of the schools in Toronto and he obviously wanted to get into film, so he contacted me; I met up with him at a party . . .

°fi°fi°fi°fi°fi°fiSo Stephen was my film reviewer. Next thing you know, here's Tony. In addition to his duties as an employee of Channel 47, [he was] on the *Black World* show, doing an insert on something or some event he had seen downtown, something he had arranged to cover. He would independently source out these events, come back, edit it, and just present it to me . . . because [Tony] had that technical skill, [it] was such that I'd just accept whatever he told me he had for next week's show. Like if there was an event happening at the Harriet Tubman Centre or the intramural football game between two black groups in Toronto, he would cover it. He would come to

The Early Years

the studio when the show was being taped — outside of his regular work hours — and I'd have him come join me on the set, maybe talk about what's going on of interest to youth in Toronto. So he would introduce the clips that he had shot by himself, so in that sense he was pioneering the concept that is very current now, the videographer thing that became a part of the '80s and '90s, where one person with a portable camera on your shoulder walks around and does the live "Eye on Toronto" kind of thing.

°fi°fi°fi°fi°fi°fiTony would leave messages on my voicemail, expressing his willingness to volunteer on the show and his persistence paid off . . . he was actually working in the building for Channel 47 [at the time] so [it was] a matter of him now taking time away from his normal duties, away from doing shoots from the Yugoslavian show, the Croatian show, Spanish show . . . [to get] the opportunity of getting to work on stuff he really wanted to. Stuff close to his heart culturally, personally, aesthetically . . . It [was] the black show that he wanted to contribute to. It was just very clear that this guy was going to go wherever he wanted to. He had a mindset of making this a career back then.

After working at CFMT for a while, I realized that I had to move on. My brother had landed a gig at this new video network called MuchMusic. He implored me to apply for a gig as a VTR operator. At this time, as much as I loved my family, there was no way I wanted to work in the same building as my big brother. In my mind,

Much Master T

I needed my own identity and path. I tried to avoid this conflict for months on end, but by then Basil had already hooked up an appointment for me to meet with Phil Dinan, the head of staff operations. I was interviewed and I got the job. At this time, MuchMusic was still in the early stages of construction; the Much building hadn't even been built yet. Back then, Canadians might have heard about MTV in America, but nobody had really seen this 24-hour Canadian video network. Looking back, I can honestly say that I never realized how much the Much building would mean to pop culture in this country. I just looked at my new VTR job as a gig at this new station that nobody had ever heard of. I gave my notice at CFMT and they had a champagne-filled going away party for me. I got piss drunk as a way of saying thanks for the memories.

CLEFT PALATE/HARELIP

Many people don't know this, but I was born with a birth defect called a harelip. This might be a cliché, but it's true: kids can be cruel. So I was sometimes teased mercilessly as a pre-pubescent. When you're a young un' you're not even sure what a defect is, or what it means. You just sense that something is different. For me, being born with a harelip meant that my defect was right out there in the open for everyone to make fun of. Harelip is the most common birth defect that occurs, but you can't tell that to kids, they couldn't care less, and most of them don't have the cognitive skills to digest this heavy piece of information. Honestly, the harelip teasing that I had to endure in

The Early Years

public school wasn't much better than everybody's other fave choice at the time, calling me "nigger." Rude kids would either say, "Hey, you have funny lips," or they'd artificially inflate their lips in an upward motion to mimic mine. When they weren't busy doing that, they would talk to me in a nasal tone, to let me know that they thought I was challenged in some way. But a kid can only take so much, so I would get into my fair share of fights. After a while it really began to piss me off.

My mother was instrumental in carrying me through this period of my life and for that I am eternally grateful. Right from when I was a baby she refused to treat me any differently because of my defect. In fact, her nursing background allowed her to go against the nursing protocol where birthing and nurturing a harelip baby are concerned. Like for example, if you had a harelip baby, women were instructed not to breastfeed their children back in the '60s. But I was a breastfed child. Despite my minor shortcomings, my mother treated me just like she would have treated any other little snot-nosed kid from Leeds. Most importantly, she talked to me, exposing me to normal patterns of speech and helping me to overcome what might have become a speech impediment.

You see, I had a cleft palate as well as a harelip. For those of you who don't know, clefts affect the soft palate, which is the posterior part of the roof of the mouth, and this in turn affects speech. If you move your tongue along the roof of your mouth from front to back, you will notice that the roof of your mouth becomes soft as you move your tongue towards the back. The soft palate moves when you speak, to prevent air from escaping into your nose and causing a hyper-nasal quality. When that happens, it becomes difficult for a child with a cleft palate to produce many of the sounds necessary for

Much Master T

speaking English. It was this type of nuanced explanation — along with daily reinforcements designed to make me realize that I was no different than other kids — that prevented me from seeing my defect as a major issue. Naturally, my father's behaviour was very much in line with my mother's. For him it was just business as usual. "Cleft palate, schmeft palate" was his motto. My brother Basil was probably the most protective of me. He would be willing to wallop anyone with a knuckle sandwich (knuckle sandwiches, not knives or guns, were big back then) who messed with me.

    I had my first surgery to begin the process of correcting the defect in England and I still remember the doctor's name, Dr. Eastwood. Clefts are usually repaired within the first year of life and doctors usually wanna operate on you when you're very young. From what my family tells me, I had between two and three operations before I was even a year old, for the cleft palate as well as for the lip. At the minimum, one surgery is needed to repair the lip and a separate surgery is needed to repair the palate. However, several surgeries are needed to make the lip appear as normal as possible. And sometimes additional surgeries involving the palate are needed to improve speech. When doctors perform surgery on the lip portion they have to entirely close the lip first and then they have to continue the surgery in the areas around the lip. As I got older and I stumbled through the awkward teen years, the concern was not so much how I looked, but that I was speaking with this really nasal vocal tone. By the time I hit the age of 16, I had already done some hard work to correct that, in addition to taking speech therapy classes. I remember how the doctors would work with me on words like "sixpence," which with my ailment would come out sounding like "thithpence." After a while I just got tired of saying "thithpence"

The Early Years

(or should I say "sixpence") and simply told the doctor "no."

As I moved into my late teens and early 20s, the harelip/cleft palate situation was always in the back of my mind. I guess you could say it kinda made me feel insecure about my looks because I always looked at my brother Basil as being the perfect handsome brother, the Young family babe magnet. Sometimes I thought maybe the chicks wouldn't dig me much, if at all, because I had a defect. However, some kind of divine intervention kicked in miraculously just before my 20th birthday. That's when I started to realize that my so-called affliction had actually given me the ability to bounce back in this game of life by teaching me to be resilient. As a selling point, it made me different from the norm and it forced me to focus on other parts of my make-up. To this day my wife Paula says having a harelip is one of the things that's shaped my strong personality. My mother seems to think that I've always enjoyed acting like a clown because of my emotional response to these defects. It's funny, my mother still delights in my off-the-wall comedic sensibilities, and laughs from the gut about the littlest things I do both on and off the air.

One thing I can say, I've never let my harelip hinder my progess with the ladies. Sure, I had one lean year in college when my confidence was down but I was still able to charm Paula off her feet. I've always thought to myself, "Why shouldn't Pam Grier want a piece of this?" And as a teenager I've always been able to get the girls too, dammit. When I was first learning to live with my defect, I liked to think I was unstoppable in the babe department. Like Shaft, Erik Estrada from *CHiPs*, and Billy Dee Williams. (Okay, so I wasn't as proficient as Billy Dee, but work with me here.) Still, this confidence with the ladies didn't always find its way into the other areas of my life. When I graduated from Mohawk College, it was a big

Much Master T

thing for me to accept that, because of my insecurities about my harelip, being a TV host might not be a career option. And I was okay with that. I convinced myself that my dream was always to be a producer anyway.

Do you remember that actor named Stacey Keach, famous for his role as hard-edged detective Mike Hammer and infamous for his drug bust and prison sentence? Well, he was my role model. He had a harelip. Sure, he was white, middle-aged, and chubby and I wasn't, but he made all things seem attainable by virtue of his defect. He's appeared on Broadway and in Pulitzer Prize-winning plays like "The Kentucky Cycle" and he's the honorary chairman of the American Cleft Palate Foundation. Hey, we're all in the same gang, I surmised. Being the burgeoning master thespian that I thought I was, when I used to audition for parts, I would stop myself in my tracks because of my lip insecurity. It got ridiculous. I would attribute the fact that I didn't get an "extra" or small part in some fluff, sub-par, B-movie to my lip. Thankfully, Paula would always pick me up by my bootstraps and tell me to smarten up and shift the focus away from my lip. Repeatedly, she insisted that I think about the content, the script, and what I have to offer the audience by way of my personality. She always felt that my auditions would be high calibre if I put my heart into it. "Hare lip, gare schlip, nobody will give a hoot about that if you're good," she insisted. Paula ingrained in me that looks are only part of what makes a successful actor; if someone's attractive and there's no content to back that up, no one will give a crap. She can be a tough cookie when she's ready.

Certainly, when the opportunity came up to be on the air at Much I was a little nervous about being judged on my looks or my appearance. But only a fool wouldn't be excited by the chance to

The Early Years

work in such a cool, visual medium. All that cajoling must have kicked in because when the time came to put together something together at Much, I wasn't as intimidated anymore. That's not to say I didn't have my moments. In fact, during the taping of my third *X-Tendamix* show, Paula and I had one big blow-out discussion about my insecurities. At one point I literally refused to go on air or to do my next show. I had somehow convinced myself that maybe I was just some loser from Hicksville, Ontario, and I didn't really deserve any of this. The question I had to replay in my mind is this: Are decent people more interested in my work ethic, chutzpah, and drive to succeed, than my lip? And is hair colour, nose size, and style of make-up a primary or a secondary concern to our viewers out there in TV land? I somehow convinced myself that these issues were trite, trivial, and moot.

When I was old enough to grow a moustache, like at 28 to 30 years old (I'm serious here, I was a late bloomer), I did. For the harelip crowd, sometimes growing a moustache can be a bit of a security blanket to cover up some of the scarring. In fact, I sported this thick, black, woolly blanket of a moustache for 10 years on the air and no one ever knew exactly why. They probably thought I was wearing it for aesthetic reasons — and that was certainly part of it. But I also wore it in a Rasta sense. If you look at some of the later *Da Mix* shows, I was also sporting a small beard and sideburns. I just let it all grow natural. To this day, whenever I mention this cleft palate topic, most people tell me they didn't even know I had a harelip or cleft palate.

But undoubtedly, some folks have noticed. One of the most beautiful things for me is to be stopped by parents of children who have this birth defect. They're just amazed at my resilience. They ask

Much Master T

me questions about how I've overcome this barrier and I always take the time to let them know that they just have to treat their child the way they would treat any other child who doesn't have a harelip or a cleft palate and the kid will do just fine. I guess the message my mom sent to me as a kid has stuck. Admittedly, Paula and I did have a slight concern that our son might possibly have a harelip/cleft palate due to genetic factors. You have to worry about these things as a parent sometimes, because more severe cases of harelip/cleft palate can grossly disfigure a person and leave them with a speech impediment that's nearly insurmountable. So I've been really lucky and I give thanks every day.

Interestingly, one day when I was taking my son for a check-up at the Toronto Sick Children's Hospital in 1999, one of the doctors came up to me and asked if I would participate in a conference to discuss this subject and invite children with cleft palate to join in the symposium. One of the conference organizers, a nurse whose name slips my mind, called me up to talk about the project. Based on how she approached me, it seemed like she was trying to determine if the topic was too heavy, too personal, for me to talk about in public. Anyway, I told her I'd be honoured to speak to the kids, but for some strange reason it didn't happen. In retrospect, it was an event that I was really looking forward to being a part of — even if it was a Catch-22. I have to admit that I was a little apprehensive about looking deep down and bringing up all the self-esteem issues I had been forced to confront in the past. Somewhere in the dark recesses of my brain, I didn't want to dwell on anybody's shortcomings — especially my own. I just wanted to be judged on the merit of my work. At this time in my life, I was very proud of myself. I figured, here I am, a black man with dreadlocks who's been across this vast

The Early Years

country of ours, getting compensated well for doing what he loves to do. And I have a harelip/cleft palate. My so-called defect had become a complete afterthought by this time in my life. As somebody with a so-called defect who's done quite well for himself, I'd still like to take my inspirational stories on the road for harelip/cleft palate children and parents to hear them. That's what is important to me: showing kids that, with the right kind of support, you can overcome anything that you come up against. My message is simple: If this small-time, cleft palate dude from Kitchener can do it, there's no reason you can't do it too. I truly believe you can move forward in life if you believe in yourself and your "calling" and you understand that there's a higher power at work. I'm not just in this profession to simply enjoy all of the accoutrements of being a VJ (like meeting A-list music stars and raking in the bucks), I'm genuinely in this to try and affect change from my heart. When the intention is pure, anything is possible.

I suppose saying things like "anything is possible" makes me sound like an eternal optimist, but my cleft palate issues haven't always been peaches and cream. Honestly, when I was around the age of 24, I started to have serious problems with my nasal passage. I consulted my family doctor, Dr. Trapp, and he referred me to a hospital in London, Ontario. In an unbelievable twist of fate, the doctor in London who wound up performing surgery on me had trained under my first doctor, Dr. Eastwood, in London, England. It was like I had gone 360 degrees in life. This last surgery in 1985 was conducted to completely clear out my nasal passages. The doctor had to do some work on lifting my nose up because it had become very flat. In fact, when I ran into a friend from high school that I hadn't seen in a couple years, he had the nerve to ask me if I

Much Master T

had had my nose broken or if I had been punched in the nose. He said I was starting to look like Sammy Davis Jr. Apparently, the bridge of my nose had flattened out quite a bit; I also had to have a bit of my lip trimmed off because it had flapped over. With a brand new sniffer and a more refined pair of lips, I was good to go. Some kids with a harelip aren't so lucky. Studies show that children with clefts are at a slightly increased risk of developing learning disabilities. To date, I've had about five major surgeries. But talking about this condition doesn't faze me now. I give thanks every day that I've been able to continue on and live a productive life.

Life at Much

ON-AIR ID PHENOM

When I came over to MuchMusic I had to re-orient myself to this new Beta system. Much had both operating systems (¾ inch and Beta), but Beta was the new thing and I was worried about my ability to comprehend this new technology. All of this was going on in a frenzied atmosphere of folks scrambling around to ensure that this new network got on the air without a hitch. The network launch date was around the same time of the year that I originally came to Canada (either August 14 or 16), so it stands out in my mind. Much launched on August 16, 1984. The original vjs were J.D. Roberts, Erica Ehm, Michael Williams, Christopher Ward, and Jeanne Beker.

Early on, I became one of those cultish, behind-the-scenes characters that Much camera operators would take great delight in showing on TV — even if it was usually just the back of my head or the occasional shot of my face. Once they got to know me, the VJs and camera operators realized that I had quite the personality, so they'd include me in all of these short vignettes. The whole thing was all trial and error and there weren't any rules, so I'd do my regular nine to five shift and then go back in during the wee hours and work on these Christopher Ward skits, sometimes staying at the building until 2 a.m. I was finally starting to feel like my life had more meaning now. Slowly but surely, I was also becoming way more well-rounded as far as music interests go. As a VTR operator I used to listen to the music on my headset and I'd crank the volume up; it exposed me to all of these other genres. Paula and I used to go to every free show as well, regardless of genre. This is where I cut my teeth as far as having an appreciation for all types of music.

MUCHMUSIC GROOVE

Here's the million-dollar Much trivia question for the ages: What is the first Canadian rap video listed in the video archives at MuchMusic? Is it by Maestro? Naaah. Michie Mee? Nope. How about Rumble and Strong? Try again. If you answered Master T and the Super Hip Three, you are correct. But before you pick up the phone and call the Nation's Music Station to request the video, remember that I recorded this Much promotional number during the mid-'80s when cheese was in. Now, I didn't exactly spark off a music video revolution with this song, but it was an important contribution to MuchMusic, nevertheless.

Life at Much

Not that I was the only VJ who made a difference at Much. Moses Znaimer has always had an eye for talent, so every VJ to step through the doors at 299 Queen Street West has possessed something original. From Dan Gallagher to Christopher Ward to J.D. Roberts to Ziggy Lorenc to Monika Deol, we all had very distinct personalities and different ideas of what we wanted to bring to the station. Fortunately, Moses had an open-door policy that allowed him to take advantage of the talent on hand. Staffers could come in and talk shop, maybe pitch ideas, or simply chat about whatever happened to be on their minds. I can't exactly remember the first time I took him up on his open-door policy and entered the visionary world of Moses Znaimer, but I do remember that one of the first meetings I had with him revolved around a MuchMusic promotional item that I had worked on with Gord McWatters. Gord and I were both lowly camera dudes in 1988, but we wanted to keep our creative juices flowing — and possibly make Canadian music video history in the process — by creating this video themed around Much moving from 99 Queen Street to 299 Queen Street West. McWatters (I used to call him Dutch) and I had good chemistry and we shared a track record of producing several well received promotional IDs for Much. The camera crews had these official "field days" where every other Friday we'd set out to come up with a concept and shoot something. It could have been a Much ID or a bumper. Generally speaking, a lot of the other camera crews didn't end up doing these pieces because it was kinda optional and if there wasn't anything scheduled to do you could sorta take the day off. Well, Gord was a really creative guy, a genius at generating cutting edge concepts (he's now an artistic director at Space: The Imagination Station). In the late '80s, Gord would come up with all

Much Master T

these ideas; he was like a little Hitchcock. I would be his leading lackey, the principal actor who'd carry out all of these wacky ideas and pull these stunts off. Remember, I was a frustrated actor who couldn't get any gigs then, so these skits were stimulating.

The first project we worked on together was unofficially entitled "The Shower." It was a take-off of the famous shower scene at the Norman Bates motel in *Psycho*. It starts off with me in the shower singing Eddie Murphy's "Party All The Time." Then there's this crescendo-like build-up of creepy music, and instead of Norman Bates coming to stab me, there's this MuchMusic "M" prop that attacks me in the shower. It's pretty comical and wacky, very reminiscent of the original *City Limits* style. Through this work with Gord, I also made a name for myself as the "Spy" character in so many Much IDs. I was the guy in the trench coat who'd run around the city with this Much "M" prop in my briefcase; I'd run through tents and in back alleys, anywhere I could brand the "M" logo. Pretty quickly, I came to be strongly associated with this character at Much. People on the street even knew me as the "Spy dude." So both Gord and I decided that these Fridays weren't just going to be another day off. We viewed the free time as an opportunity to work through some ideas. I know my love of the work had everything to do with this. I could have just taken the time off, but I took these days as an extra opportunity to make a dent in the public consciousness. I've never been afraid to go that extra mile, and I think that's why I've had relative success in this biz. Whether there was any moolah involved or not, I would spend dozens of hours plying my actorly trade just for the sake of honing the craft. I used to always rush home to Paula and tell her how much I loved these little projects. It was all about experimenting with the acting process, something I've been interested in throughout my

Life at Much

career. It's amazing how I remained passionate about these IDs for four straight years. I'd always put the viewer and the final product ahead of anything else — food, sleep, soccer practice . . . well maybe not soccer.

As our IDs started to generate this little street buzz, I would occasionally be stopped downtown by folks who'd say, "Hey, you're the guy that does the spy or psycho thing." The spy IDs unexpectedly gave me a cult following, which was kinda cool. It gave my acting shtick some more leverage and kept me in the public eye. Around that time, Kim Saltarski, who had gotten me the stunt double gig on *Police Academy 3*, came up with this brainchild script for a short film. He described this situation in which a cow prepares to go to space, and I was to play this pretentious little model. I was experimental to the nines by now and so we drove out to Westover, Ontario, to shoot the project over at his parents' farm. When we were shooting this bit in Westover, we set up in this field where all these cows were lying around. Now, if you know anything about me, you know that I hate mud, manure, and wet environments. So you can imagine my reaction when I looked up to see a cow pissing right in front of me; there was another one taking a crap right in front of me too. My character was supposed to be this pompous, uppity model who didn't want to have anything to do with the slime and slop. Decked out in an '80s version of a funky, spacey jumpsuit (accented by black leather, a '60s-style belt, a pair of vintage black cowboy boots, and an earring dangling from my left lobe), I was certainly looking the part. But there was no method acting or roleplay involved here. This was as real as it could get, and I wanted out.

The funny thing about this shoot (aside from the manure and the ridiculous outfit) is that it took place after Paula and I had gone out until 6 a.m. the night before. I had been out dancing for eight

Much Master T

hours, and I was pretty fatigued. Still I made it all the way out to Westover to stand in manure all day, for no pay. Now that was paying dues. Paula was shocked that I had managed to get up in time for the casting call and she couldn't believe that I was still going. She always reminds me how I'd do all of these gigs that people in the biz wouldn't piss on, and it would always be fruitful in the end (to this day, people still ask me about the *Don't Look, Cow* short). Anyway when it was completed and we went to the first screening of *Don't Look, Cow* at the old Opera House, no real audience turned out; it was just friends and family. We all watched it, agreed it had cult classic potential, and that was the end of it. Or so we thought. A short while later, Kim called me, out of the blue, to tell me that he had sold the spot to First Choice (an '80s cable network that used to have pay-per-view movies 24-hours a day). They would play a lot of interesting short films between feature movies — and they used to play the heck out of them — so, the fine folks at First Choice (and Kim, of course) were truly responsible for boosting my mini-star status at the time.

After doing the ID with Dutch that promoted the move from 99 Queen Street to 299 Queen Street, a lot the folks at Much were hot on us to produce a full-length video of sorts. But how were we going to pull this off? A full-length video — with no money — while we were working full-time? One of the answers came in the form of a band that played at Much. It was really just a bunch of fellow co-workers who liked to get together and jam on their days off. Now, Paula had written a couple of rap sequences that she had performed with a local DJ in Kitchener when she was about 15 years old, so the two of us figured that we could come up with something creative. The video itself would incorporate all of the VJ personalities and

Life at Much

shows on Much, as well as behind-the-scenes staff. To pull this off I had to hook my boys up and convince them that this was gonna be a hot project. Gord and I decided to give them a bunch of silly aliases and I called 'em Master T and The Super Hip Three. The band consisted of Steve Vogt (a.k.a. Steve Snare), Dave Murphy on guitar (Lo Tide), and Richard Oulton on bass (Richie Baby). I got Paula (Lady P) to be a co-lead vocalist and sing back-up, my brother Basil would be our DJ (Mix Master Baz), and Gord was our shifty manager (Dutch, as I mentioned before). Paula and I came up with the concept and named the band the Super Hip Three.

At the time, we lived together in a tiny bachelor apartment. It had a big waterbed in the middle, a sewing machine on one side, the TV beside that, and a small dining area. We used to call it the Cracker Jack box, because it was such a tiny place and it gave life to so many of the inspired ideas that were flowing through us at the time. I learned how to rap there, Paula learned how to write songs there, and "The MuchMusic Groove" set things off for both of us. After Gord and I had painstakingly come up with this plan to execute the production, I set up a meeting with Moses to talk about it and get his okay — which we needed in order to fund the project. If there's one thing most staffers knew about Moses, it's that he consistently had his ear to the ground, so he had already heard about our plan. The best time to meet with Moses about anything was in the late evening. So I went to his office around 8 or 9 p.m. and knocked on his door. I explained to him that I was interested in doing this video and I asked him to give a listen to the demo we had already done. I left it with him and he gave it a listen. We talked in the hall one day and he gave the project the go-ahead. With the exception of a few small changes we had to make to some of the vocals, everything was ready to go. I went back to the studio to make things right.

God bless Moses for giving us the green light to fulfill our cheesy '80s video dreams. We laid down the track at a studio owned by these two British guys, Mark and Steve (later of Bookroom fame). But the best part about the video had to be the footage. My outfit was reminiscent of what breakdancers used to wear: a nylon track suit, a baseball hat, runners, sunglasses, and a fake gold chain — Run-DMC style all the way. The rest of the band wore jeans, leather jackets, runners, and sunglasses. As I said, the video incorporated everything that made Much what it was at the time (including all the VJs, their shows, and the building staff, among other things), so the shooting of the video itself was shrouded in secrecy. We even incorporated some rare footage of Moses dancing in his office that appeared at the end of the video.

We created a bio for the band and in the bio we used all the fictitious names of the band members; Basil shot some great pictures that made it look like we were actually a real group. As much as we enjoyed producing the video, the brightest moment for me and Gord was yet to come. We had planned to hold a launch party, upstairs at the Beverly Tavern just across the street from Much. But we couldn't really spread the word yet because we were still trying hook up a budget in order to get some food and other expenses taken care of for the party. So even the launch was a bit of a secret.

As planned, we launched the video at the Beverly and the place was packed. It was almost like a movie premiere that night. We blew our $400 budget on wings and finger food and everyone had a blast. "The MuchMusic Groove" was well received by all that night, but the real beauty of it is that it injected the staff with a new sense of morale and togetherness. It also wound up being an incredible

Life at Much

promotional tool for Much. And to think it was all done for free.

The day after the launch, Much immediately put the song into what they call B rotation, which meant that they spun it once or twice a day. As you all know by now, Gord and I ended up making history here because this was the first Canadian hip hop video to enter into Much's library. I know it's a frightening thought, but there were virtually no rap videos being submitted at the time. After the success of "The MuchMusic Groove" video, we heard through inside sources that it was being used as a promotional and educational tool to be screened whenever business associates came by or when tours came through Much. We were told that, from a conceptual standpoint, this video was over and above anything being put out by the promotional department at the time.

Now, it wasn't all peaches and cream for me. Not everyone understood the angle I was coming from in terms of mixing comedy and promotion. Some people were still questioning who I was — I mean, I wasn't an on-air VJ, so what was the deal with me anyway? For the ones who got it early on, great! For the ones who didn't get it right away but caught on much later, what took you so long?

"The MuchMusic Groove" lyrics (1988)

Master T is back with a heavy groove. Don't touch that dial, don't nobody move. There's a few things I've just got to say, about the great shows MuchMusic brings your way.

Well the week starts off on the mellow side, with Ziggy's *Mush Music*, the Lover's Guide. With my satellite dish I point

Much Master T

to the west to catch TDM with the West coast's best. Oldies, goldies, videos from the racks. TDM also brings you *BackTrax*, the Stones, Madonna, and a whole lot more. Get to strut your stuff on the spotlight floor.

Get on up and let's move to the MuchMusic Groove. x 2

And now for videos that are tasty and new, Much and the Munchies got something for you. Hostess the mostess, the new video show, the latest hot videos in the land you know. Well it's time to turn your speakers to overdrive, 'cause the *Pepsi Power Hour* is coming alive.

Get on up and let's move to the MuchMusic Groove. x 2

There's just no missing the *Coca-Cola Countdown* for the Top 30 videos. Around the town, look everybody, here's a picture of me, I think I'll mail it in to RSVP. Requested songs for video play, could I see Michael Jackson today?

For style, fashion, and so much more, tune into Erica Ehm for *Fashion Notes*; scoops, stars, musical views, Kim Clarke Champniss has all the news. He keeps you in tune with an hourly rock flash, he has a crack news team with such panache.

Get on up and let's move to the MuchMusic Groove. x 3

Soul in the City is a funky show, *Soul in the City*, soul music on the go. *Soul in the City*, to get you off your feet, *Soul in

Life at Much

the City, Michael Williams gets the beat. On a Saturday night, don't leave your pad, 'cause *The Big Ticket Special* is super bad. Concerts, specials to blow your mind, and you can listen to your radio at the same time. New media revolution, what's that you say, *City Limits* with Chris will show you the way. For offbeat videos, independent bands too, tune into the show that's innovative and new.

Get on up and let's move to the MuchMusic Groove. x 3

So as you can see MuchMusic provides 24 hours of musical pride, shows with romance, new soul, and power. Make sure you tune in every hour. I'm Master T, and the Super Hip Three. We hope you enjoyed this musical odyssey.

Although I hadn't yet done an audition tape, "The MuchMusic Groove" shtick got so many people thinking that I wanted to be a vj. But they were so wrong. I still just wanted to be a producer because, in my head, a vj wasn't an actor and vjs didn't usually do character roles as part of the job. Just because it would have meant that I'd have been in front of the camera didn't mean that I would have been able to experience what I wanted in terms of performing. Plus, I had my own fears about the business and I was concerned that it might change my life in a way that I wasn't ready for. Now, when a position came up as a floor director on *Electric Circus* (hosted by Monika Deol and Michael Williams at the time), that was different. I was really gung-ho for the job, which made things even harder when I didn't get it. At this point, I figured that the on-air thing wasn't looking so bad after all. I knew it would be

Much Master T

somewhat uncomfortable for me, but I also knew that I could be more creative if I had full control of the product and my image. So when I got a hot tip that a three-hour time slot was going to be devoted to playing dance music (which was very popular at the time), and that the show didn't yet have a host, I knew the Creator had bigger plans for me after all.

My personal life, at this time, was blossoming too. I had been with Paula for approximately six-and-a-half years and we were still going strong. I finally decided to propose to her so we could make things official. After a long engagement, we finally got married on August 19, 1989. When we first began planning the Event to End All Events (that was what the first line of the ticket-style invitation entitled Matrimonial Interlude read), we wanted to create an unforgettable experience for our guests. We utilized the services of a three-piece jazz band headed up by popular local musicians Dave Williams and Neil Brathwaite; they played standards from the jazz, funk, and R&B genres. DJ Lenny from Kitchener (now a popular DJ at Toronto's Club Paradise) supplied the wicked house music, reggae, and hip hop grooves. The special song that I wanted the DJ to play was a Luther Vandross number, but we went with the band's and Paula's suggestion, Sade's "Your Love Is King" instead. The ceremony and reception took place at Toronto's Great Hall and the event was catered by the Real Jerk. Paula designed and made most of the clothing for the wedding party. My best men were my dad and Basil, and the maids of honour were Paula's sisters, Donna and Allison, with her brother Kirk walking her down the aisle. The wedding was also a special moment because my dad came in from New York; it was his first time in Canada. Even Brona (host of *Brona's Video Diary* on City TV) came by and taped the wedding for

Life at Much

her segment, so you know it had to be something special. To this day people still mention how incredible that day was.

X-TENDAMIX

Moses always thought I wanted to be in front of the camera — especially after the success of "The MuchMusic Groove" — but I still wasn't sure. Only after three separate meetings with him did I finally agree to do my own show. And even then, I had such fears of being on the air that I had to come up with an on-air character for my show. I just couldn't see myself as a VJ before this time. I just wasn't ready.

Moses Znaimer
president of MuchMusic

The most significant thing about my impact on T's life and his career is that I insisted, and he rejected. I proposed, and he said no. I suggested, and he was kind of vague. In other words, I was acquainted with T, I was acquainted with his brother, and I knew them in the roles that they were occupying at the time. When we decided that we were going to get more heavily into urban music and what eventually became *Da Mix*, *Rap City*, and so on, I started looking around my own building and I thought T was the guy who looked like the casting that I wanted to do. And having spoken to him a little bit, I felt certain that he was the right choice. So I call[ed] him up here one day and laid it on him — and normally people

Much Master T

fall all over themselves, or they're beating on the door on their own for months and years — and here I was offering T this gig and basically he said no.

At the time I thought, "well, there's an unusual thing," and "my, what modesty," and the truth is I, to this day, am not certain whether it was self-doubt or whether he had thought it through and just didn't want that kind of public life. For example, his brother Basil is equally talented, but wasn't as interested in the celebrity; he had a more linear focus and that was to be behind the scenes, and maybe T was the same — he was brilliantly cagey. I actually had to pitch to him three times and I'm certain he was the right guy for the gig after making two offers to him. Y'know, there are some who can't take yes for an answer.

When the opportunity to be involved with something that I could have full creative control over presented itself, I knew it was time to change my thinking and go for it. For me now, it had become quite clear. Everything was pointing towards being on the air. I talked to John Martin, who was the Director of Music Programming at the time, and got an insight into how the whole dance music thing would play out once it hit the airwaves. I called Paula and we went to work. I figured that if I did a pilot and visually showed Moses and John what I had come up with for this host-free three hours of dance music, maybe I would be given a chance to develop my own show. Paula and I put together a pilot for *X-Tendamix* and Moses liked the concept. Exactly two years after the success of "The MuchMusic Groove," I was ready to have my own show. The only catch was that I would still

Life at Much

have to do my camera job, part-time, if I really wanted this chance.

At this point in my life, the biggest fear I had about going on the air was that I didn't want "the biz" to penetrate my character, get to my head, or affect my relations with Paula. I wanted to be as far removed from the showbiz industry spectacle as possible. So I figured: what better way to do that than to develop a made-for-TV split personality. *X-Tendamix* was my first show so I wanted to come up with something that was animated and bursting with life.

After much introspection I came with the name Master T. The name came from the character I had made up for Master T and the Super Hip Three on "The MuchMusic Groove" and people kind of knew me from that. So I spent many an hour developing this character. Master T was over the top, in-your-face, and funky. There was also another character that was closely associated with the show right from the beginning. Her name was Roxy, and she was a keyboard sampler. Roxy was conceived during the time the original pilot was shot. I had come up with the idea of having a sidekick so that I could interact with another energy to make the throws more exciting. When I first told my wife about the idea of having a talking sampler on the show she didn't quite get it, but she's from the school of 'you don't know 'til you try,' and she agreed to be the voice of Roxy. All the banter between myself and Roxy had to be prearranged and all that dialogue had to be thought out and recorded every night before *X-Tendamix* was to air. Roxy became a well-loved element of the show and she even got letters and special requests from viewers quite often (including fan mail from incarcerated viewers who wrote to say that she lifted their spirits with her sexy voice). I would get fans asking for her and yelling her name out on the street. Unfortunately, when Paula became pregnant the second time she was unavailable to

Much Master T

do the voice any longer; Roxy was unceremoniously phased out.

X-Tendamix was to be a high-energy dance show that played blends of videos from hip hop and reggae to house music and soul. The concept behind *X-Tendamix* was quite simple: our playlists were set up to represent popular demand. This is where my DJ, Dave Campbell, came in.

Dave Campbell and I had met years earlier at the infamous Twilight Zone after-hours club. This was where Paula and I first heard D.C. spin tunes. His skills stood out to us ever since that day. Consequently, a few years after the club closed and I knew that *X-Tendamix* could become a reality, he was the first person I thought of for the job. Once we made our musical connection, Dave would come down to the show and hip us to all the freshest new grooves. This was perfect because my vision was to create a constant wall of sound throughout the entire three hours of the show. So if I was doing a throw, or if there was a performance or maybe an interview, there would always be music playing in the background to promote an uninterrupted flow of the latest tracks. His knowledge of music and his dedication (he would show up every week, select the tracks, drop them off, and pick them up, all for no pay back then) impressed me.

My profile at Much was growing quickly. I was only involved with *X-Tendamix*, but it was a lot of work. I'm not at all convinced that there's any place that would have given me the amount of freedom and power to produce my own show in the no-holds-barred format that I was given. Without creative control, the show could easily have turned out to be lame, boring, and bland (like some things you see on Canadian television), but by virtue of the fact that it was on Much, it really couldn't have turned out this way.

In the very beginning of *X-Tendamix* I decided that I wanted the

Life at Much

show to have a hip and funky set. So I came up with this cool idea to put a clothesline up behind me in the studio as a backdrop to display message-oriented T-shirts. It was a subliminal way for me to get positive cultural and political messages out to the viewers and stimulate their consciousnesses. This clothesline became very popular and soon a lot of small T-shirt companies were contacting me so they could use this creative vehicle to promote their product. As a result, viewers would call me up to find out where they could get some of these amazing shirts. In due time, Paula came up with the idea to open a store that would house all these different shirts from independent companies. She also brought in some hot clothing lines from the States like Cross Colors and A.A.C.A. With her design background, she managed to create the perfect mix; the store also gave her an outlet for her custom work. We named the store T's Cribb and away we went on another fulfilling endeavour. Although the store had my name on it, my involvement was minimal because my plate was already so full.

Now, *X-Tendamix* was not some haphazardly conceived hodge-podge of tracks from every genre imaginable. It was a carefully executed plan to show the full range of black music that had dominated so many people's lives, especially genres like reggae, house, and slow jams that were being neglected on both commercial radio airwaves and on TV as well. It was a three-hour show that was broadcast twice every Saturday.

Between the time I completed the pilot and the time the show actually went to air in September of 1990, six months had passed. It was also in October of that year that we were able to confirm that Paula was pregnant with our first child. So many things were going on at this time in my life. Paula was feeling great, and we were in the

Much Master T

process of packing for our big move out of the Cracker Jack Box into a three-bedroom loft-style apartment that was also in Cabbagetown. There were so many new beginnings on the horizon for us back then — except when I thought about my Uncle Rupy. We had known for a few months that he wasn't well, but we didn't know to what extent. The cancer he had been diagnosed with had begun to spread more rapidly just as my first few *X-Tendamix* shows were going to air.

When I did my first show, it was almost surreal for me. Could this possibly be happening to me after all this time? It most definitely was, and the reaction from the public was truly unbelievable. Most people couldn't believe what they were actually hearing and seeing: dance, R&B, hip hop, reggae, and soca music in succession for three hours. It was a well overdue program. Looking at it from a musical perspective alone, it was clear that urban music fans were craving black culture from MuchMusic. We were the only Canadian avenue through which a lot of this music was being broadcast.

In between hosting and producing the show, doing my camera job, and making sure all was going well with Paula, I would make trips down to Kitchener to see my uncle, who was now living in the hospital. It was a very emotionally charged time for me because I knew in my heart that my uncle's condition had become irreversible. But nothing would prepare me for the events that would unfold in my life as I headed into the end of November 1990. My mom, Basil, and I persevered with my Aunt Sis through the horrifying process of watching my beloved Uncle Rupy slowly deteriorate to the point where he lost his sight. Because he was ill, he had never seen the show before. So on one of my visits with him, I planned for us to watch *X-Tendamix*. Even though he couldn't see everything on the show that day, he was able to hear the special dedication I made to

Life at Much

him. I could see and feel that he was so proud of me at that moment, and I will never ever forget the look on my uncle's face. He was enjoying the music; I remember that visit in particular as one of his more vital days.

When he ultimately passed away in late November 1990, I was really crushed. With Christmas right around the corner I had to attend my very first funeral. That was such a traumatic blow to me; I pledged to keep my uncle's spirit alive through everything I did as a result of all his teachings. But one thing I didn't get the opportunity to learn from him was the art of letting go. I would have to journey into those waters all alone.

There was also more adversity to follow. One day when I was running around at work, I got a call from Paula. Her tone seemed a little odd on the phone to me. She told me that something out of the ordinary had occurred with the pregnancy and she felt that we should go to the doctor and check it out, just to be sure. I rushed home and we went to the doctor late that afternoon. Her regular doctor wasn't in that day so they told her to come in and see any doctor that was available. The doctor examined her in a nonchalant manner and told her that it was just a matter of days but the fetus would eventually abort itself. I thought to myself, this is almost the fourth month of the pregnancy, how can this be happening? Paula refused to accept the information and demanded an ultrasound. She knew in her heart that the baby was still alive. When she came out of the doctor's office that day, our whole world had completely turned upside down. She went for the ultrasound the very next morning and the news was good. The baby was fine and, by all accounts, it looked as if things could still work themselves out. She was under doctor's orders to remain in bed for the rest of the pregnancy and that's exactly what she did.

Unfortunately, a month after that incident she miscarried. It taught us both a lot about letting go, and even though it was a horrific event in our lives, it truly strengthened our bond and brought us closer together. Masking the pain of these events wasn't easy, but as in the past, I was able to draw on my comedic side to get me through the ache. I had to be thankful for what I had: Paula was okay and we could try again except for the emotional battle we were both fighting in an effort to understand exactly what went wrong. Thank God she had her own business at the time. Otherwise it would have been a huge inconvenience for her to take a hiatus from it all so she could focus on her health.

While times were tough at home, my professional life was igniting. I didn't know that the show would take off so fast, as far as our viewership was concerned and from a fiscal standpoint. Three weeks after the show was on the air, we picked up a sponsorship deal from Mars Bar candy. In addition to Mars commercials being run on our show regularly, this meant that every week we'd hold contests to give away Mars Bar prize packs. "Groove the Globe" was a contest we ran on *X-Tendamix* during the spring of 1991. It was sponsored by Mars. In tandem with Mars, I came up with a contest that would send a lucky winner and a guest on a trip to New York, Paris, Germany, and London, where they'd be given prize money to groove in some of the hottest clubs in the world. The whole trip would be videotaped and the special would be aired on *X-Tendamix*. When my show took off, it was only the beginning of a special journey I would travel with a very passionate audience.

Life at Much

VJ 101: HOW TO BE A VJ FOREVER (WITHOUT GETTING TIRED)

In my 12-year VJ career I've been called the "acme of cool" and the "epitome of urban hipness" in the now defunct (but well missed) Canadian trade magazine *The Record*.

These great editors (whom I love to death for obvious reasons) also implied that my Master T shtick deserved a comic book or an animated movie. I suppose there's a reason my fan base ran the gamut from punks and ravers to skaters, hackers, hip hoppers, and the dancehall massive. I tell everyone that the key to my success is that I kept everything loose during live broadcast interviews with the stars. Maybe this approach works because I've never considered myself a star. I definitely see myself as visible, recognizable, but I'm no star. Is there really a star system in Canada yet?

And what exactly is a VJ anyway? Are we the figureheads of Much, offering up insightful, astute commentary on the state of the video universe and the world of video production, or do we just simply introduce videos and upcoming events, casually dabbling in a bit of celeb rumour milling and innuendo? Does becoming a VJ require a firm post-secondary educational component followed by graduate work in VJ paradigms? And how important is improvisation and spontaneity in the Much mix, where co-workers and tech crews are in the backdrop of every shot on the live set, doing their thing and hurling comedic insults at you while you're going live? Much employees are always shown in the backdrop working or plotting strategies to move up Much's corporate ladder.

There are no two ways about it, from J.D. Roberts, Rick Campanelli, to Sook Yin Lee, then break it down to me, Master T

Much Master T

(hey this rhymes), we all had our own individual styles that we delivered to the nation. So how does a VJ go about developing an on-air style, anyway? I was fortunate to have come from the old school VJ camp: Christopher Ward, J.D. Roberts, Erica Ehm, and Michael Williams. The Original Four laid down the foundation for what Canadian viewers had come to expect from a VJ. Even though I was never really interested in becoming a VJ at the time, it was only natural for me to observe them in action. As a VTR operator and a cameraman I got to check them out first-hand; I learned to appreciate their strengths and weaknesses. An important VJ law I learned from the Original Four is that you always have to prepare by researching your subjects thoroughly. Some might think that goes without saying, but you have to realize that it's not just a matter of cramming. You can't think you're gonna slide through interviews with some of the bigger artists that pass before you on the hot seat without knowing your stuff. You have to take the time to digest the proper information and use the research to come up with insightful questions. It's not gonna be a good interview if the artist doesn't think you're up on your game. Another important thing I learned observing J.D. Roberts is that you can never allow anything personal to interfere with your professionalism on the air. One afternoon, J.D. had a very heated argument with another staff member just before he was scheduled to go on the air. Watching him go through this, I never thought he'd be able to compose himself after the blowout, but to my amazement he shifted gears quickly and delivered a flawless throw. It was as if nothing had happened. But right after the show, he went right back to the argument with the same intensity. That blew me away!

Personally, I used to approach my on-air shifts with the idea that

Life at Much

I'd just go with the flow. I would always try to work with whatever energy I was carrying at that moment, and I'd try not to pretend to be up and hyper unless I was really feeling that way. I think it's a more natural approach. I would also make an effort to achieve a balanced karma before I had to begin work at any level. Sitting at my desk and enjoying a tea and a cookie always helped. And when it came to interviews, my mandate was always to be able to walk away and say, "Man, I learned some new piece of information from that celeb, and my viewers did too." Pre-interview, I have a zone I go into where I isolate myself and go over my questions, strategies, etc. I try to envision how the whole thing will play out. That's really crucial for me. Whether I'm interviewing Creed, Madonna, or Puff Daddy, it's the music that's the common thread between them and I try to concentrate on that. As far as my research strategies go, I may read other peoples' interviews but I never watch anyone else do interviews. I don't want to be influenced by anyone else's interview style because I think my free-thinking approach works best for me. And I've always had a way with people. Like my pops, I'm definitely a people person. I've come to rely on my naturally open, easygoing personality to make celebs feel relaxed.

In my years at Much, it's funny, but right around the time of the national VJ search, almost every MuchMusic staff member would feel the need to lie about their place of employment. This was due to the overwhelming amount of pestering they would sustain in constantly having to give out advice and offer insight on how to land the gig as a MuchMusic VJ. So, my former associates at the Nation's Music Station, you can now breathe easy: you'll never be bugged again. But for all the wannabe VJs out there, here are my top 10 suggestions on what it takes to be a great VJ:

Much Master T

1. Personality (make sure you have one).
2. Music (immerse yourself in all genres).
3. Research (give yourself three days of leeway and make sure you know your stuff).
4. Did I say personality?
5. Get yourself a post-secondary education (not mandatory, but when the gig is up you'll need something to fall back on . . . although UPS is always hiring).
6. Resilience (develop a tough exterior to repel armchair critics).
7. Be creative (originality and fresh ideas go a long way).
8. Listen to your subject (you may love to talk, but there's so much to learn when you actually listen attentively).
9. Style (visual presentation is key, so personalize it).
10. Be yourself (easier said than done in front of a camera).

ALTER EGOS

Master T is high-energy, jittery, and zany. As for that Tony Young fellow, you won't hear much more than a peep out of him. He keeps to himself, he's humble, and he absolutely abhors loud, obnoxious types. So the Master T you've grown to love is a radical departure from Tony. In my 12 years on the air, I've had to come up with all these alter egos to keep my show's energy fresh. Would Canadian teens and twentysomethings truly have enjoyed seeing the same guy introduce videos, video clips, and artists for over a decade without getting bored out of their skulls?

I don't think so. That's why I spent so much time parading

Life at Much

around as any one of my alter egos. There was Monty T (Monty Hall game show host, minus the Hall, plus the melanin); Taurus T (Barry White with a capital T); T McGhee (the first kilt-wearing Jamaican-Scotsman on national TV); Selecta T (my dancehall loving cousin); and Santa T (the black funky version of St. Nick). Even before I got my own show on Much, I always wanted to make up characters that would fit me to a T (pun fully intended). But let me make something perfectly clear. No, I do not have a multiple personality disorder, nor do I have any form of incurable neurosis. But I do subscribe to the Sammy Davis Jr. approach to performing. And that means I'm all about entertainment to the fullest. That's how I became The Brother from Another Planet.

Growing up in England I used to do impressions of everyone from Groucho Marx to James Cagney to Louis Armstrong. At the beginning of *X-Tendamix* I wanted the viewers to be as attuned to me as they were to the videos. That's why Master T was so over the top. For the most part, I used to wear flashy club gear and I always wore shades, along with a variety of hats. I used to get a lot of clothes from a store called Rag Tag located right on Queen. They carried some of the most eclectic, high-end causal gear in downtown Toronto (the Queen Street West fashionistas seemed very impressed). But just two months into the show, Paula was getting quite concerned; she thought I was going too deep into the characters. You have to remember, I had just started to dreadlock the top of my hair, and the back and sides were shaved, so this Master T guy was an extreme looking character for TV — especially back in the early '90s.

Obviously, the first alter ego or character I conceived of was Master T. At first I was gonna call myself Much Master T, to signify someone who controls both the mic and his own destiny, but that name got

Much Master T

turfed by Tony Young at the eleventh hour. For the record, the name Master T is as shallow as it sounds. It has no double meaning, no symbolism, and it doesn't imply references to anything. It's a name I made up on a whim. Kind of the same way hip hoppers call themselves King this or Queen that. Interestingly, some folks in the black community were clearly uncomfortable with the master-slavery connotations. When you say the word "master" in the black community, even when it comes from blacks themselves, it sends shivers down our collective spine. It reminds us of a time when history was not so good to black folks. On the other hand, having white people call me "master" anything came off sounding like some strange oxymoronic reparation. White people calling me "master"? It was as if I had taken myself black to a future where things were reversed. But at the end of the day I don't think it made a difference if you were black or white. I don't think any folks from any persuasion really understood the character at the time. But they were intrigued by this off-putting, off-kilter, hip hop-styled maniac.

Monty T was conceived when Mars came along. He was like a black, old school, hip hoppy, Monty Hall-styled host with a rapid-fire delivery that made him sound like he was auctioning off the prizes. On the show's set, I would use this miniature spinning wheel as a prop that I'd have the guests rotate to win prizes. Some members of the local music community (like Carla Marshall, Devon, and Messenjah) would drop by and spin it on occasion, as my loud plaid jackets fluttered in the wind. Monty T was a sartorial clown, wearing the same type of hokey outfits that the, uhh, real Monty Hall would wear. He served his purpose and he fit the situation to a T, but when Mars left us as the sponsor, Monty said goodbye too.

With the Monty T skeletons safely packed away in the back of

Life at Much

my closet, I came up with my next character. This was T McGhee. This character was based on the accent and behaviour of my old soccer coach named Mr. McClain. You see, I had this good Scottish friend in public school named Bobby McClain and his father was the coach of our soccer team. Every time we'd be on the sidelines during the soccer games he'd be talking to us incessantly about how to score goals. I could barely understand the guy, but I had a great ear and a penchant for mimicking accents and sounds. McClain's voice stuck with me.

Growing up in England, I had frequently heard Scottish people speak, but it was even cooler hearing him talk. I can still remember so many great moments spent sitting on the sidelines listening to his accent (that was when he had his teeth in. When he didn't have his teeth in, it was a whole different story). I've always been a keen observer of unique personalities, and he was certainly one. He was also a great guy and terrific coach. So when it came time to put this Scottish character together for *X-Tendamix*, I knew I could do the accent, but how would I be able to do the character? Well, as it turned out, it wasn't all that hard to get into the role. I had this tartan vest lying around my house and Sherri Greengrass, my friend at work, found me a matching skirt in the same tartan pattern.

So T McGhee was born. He was supposed to be my annoying cousin. The type of cousin you avoid at all costs. We'd do these split screen effects to make viewers believe he was dropping by the studio out of nowhere to nitpick at his cousin, Master T. But he was fun-loving too. I could always get T McGhee to do the downright silly things that Master T wouldn't or simply couldn't do. From a technical standpoint, my production crews really worked hard to make this alter ego thing work. When not doing split screens for my

Much Master T

various cousins, the tech crew would pre-tape the lines of my alter egos and then fuse them with Master T's for show time.

In retrospect, Much was radically different back in the late '80s and early '90s than it is now. Not to put down the work of the VJs today, but we used to take so much time to work on these skits. As much as Much was about music, it was also about playing the best videos and having a good time doing it. Christopher Ward certainly has to take credit for infusing the building with this approach. His show, *City Limits*, pushed the envelope as far as skits were concerned. When Chris Ward left Much, I subconsciously decided to carry on his tradition of hamming it up for the camera at all costs and presenting these off-the-wall ideas. When you had a media visionary like Chris Ward around it challenged the production crews to be more creative, so everybody benefited.

As for what became of T McGhee, well, some members of the Scottish Canadian community in Toronto were up in arms over my characterizations. The irony of it all is that it wasn't until after I did the *Snow Job* show that I really integrated T McGhee into my repertoire. At the 1991 *Snow Job*, Master T and T McGhee had been sharing a condo together, so we pulled the split screen routine on all of these poor, unsuspecting Much viewers. After *Snow Job* I received a call from a Scotsman who was incensed by my characterization. This guy called Sarah Crawford, the current manager of communications who used to deal with complaints back in the day, and he was completely pissed off. I spoke to Sarah briefly and told her that I'd like to call the guy myself to smooth things over. I dialed his digits and he was seething. I said, "How are you doing sir, so what's the problem?" He replied with something along the lines of "I just don't think it's right, it's just not right." I said, "Uh, what's not

Life at Much

right." Then he launched into this diatribe. "Just the fact that you're up there wearing the tartan," he said, "me and my grandfather were totally appalled by it." I told him that other Scots enjoyed it, that it's just a comedic parody, and that he should lighten up. Naturally, he said, "No, no, no, no, no." I asked him if his beef with the character came from the fact that he's a black Scotsman. He denied that this was his issue. But then he asked me how I'd like it if a white person put on blackface. I said, "Uh, actually that's been done many times over. Al Jolson. Hello!" And besides, black folks live in Scotland and play on the soccer team, so I didn't see the big deal. So we spit words back and forth for a few minutes. By the end of the conversation, we obviously weren't going to see eye-to-eye and that was a shame. It was at this time that I started to re-consider doing the character. Maybe I was insulting people without realizing it. But then I got all these other calls from Scots who worked at Much and outside Much who thought the character was a riot, a hoot. So I carried on with this slapstick for a few more shows until I could come up with some other relatives for Master T, who was looking to expand his family by now anyway.

On my show I'd routinely play soft R&B love ballads because everybody needs love, right? This was the beauty of my show. I could spin any black music from the vaults, and this opened a lot of ears to the fact that there's more to my culture's musical tradition than R&B and rap. The next character that I made up, Taurus T, was perfect for bringing out some of my music's mellower textures. The idea for the character is that he was supposed to act like my older cousin who had an easy time with the ladies. Everybody has a cousin like this: the Brylcreem, the slicked back hair, the cheap suits. Taurus T was the lover boy heartthrob cousin with the deep Barry

Much Master T

White baritone drone who passed out advice to all Canadian lovers. Naturally, Taurus grew out of my love for Barry White, though he was named after my astrological sign. He was unstoppable and unflappable, a consummate disciple of love. He would come on the show to give my (mostly male) viewers help with love issues.

Taurus was a prodigious scribe who'd write poetry, sonnets. He even wrote a book of love poetry. To achieve the deep baritone rumble, my production crew used to put this effect on my voice to lower the pitch. Also, Taurus would punctuate all of his sentences with "ahh, yeah." Whenever I'd play slow jams on *X-Tendamix* I wanted this character to intro the tunes. So basically, whenever Taurus T would drop by on Master T in the studio, you just knew the slow jams would come. On my Valentine's Day show for 1993 I hooked up with a talented band of musicians that I knew from the local Toronto R&B scene. Their band was called Panic and it featured Orin Isaacs (now sidekick band leader and bass player on *The Mike Bullard Show*), Neil Brathwaite (a master sideman saxophonist), David Williams (keyboardist and producer), and Wilson (the drummer who's also on Bullard's show). We teamed up and called ourselves Taurus T and the Love Orchestra. We put on a live show for Valentine's Day, took a few phone call requests from teens who wanted to hear particular songs, and read some love letters that were mailed to the show.

The Love Orchestra performed covers of classic soul ballads by Donnie Hathaway, Stevie Wonder, and Minnie Ripperton, and I performed right alongside them. I did my own rendition of "Knockin' Boots" by H Town (this was the Taurus T remix, of course). At the end of the day this show provided great exposure for Panic, who went on to back up most of Toronto's best R&B vocalists at one time or another (everyone from Divine Earth Essence to Michelle

Life at Much

"Boo Boo" Brown). It's funny, because Love Orchestra band leader Neil called me for the next several years wondering, literally, where Taurus was hiding out on Valentine's Day. It became part of a running joke for us. This character was one of my most credible. We were able to lure the cream of the R&B vocal crop to come in and croon over Panic's brand of retro-funk grooves. Simone Denny (of Chris Sheppard's BKS fame), Wade O. Brown, and even Glenn Lewis (of Epic/Sony superstar fame) all stopped by the show. With the help of Taurus T, the segment was developing into this great way to showcase our Canadian vocalists. And my alter egos were becoming just as popular (if not more so) as Master T.

Now, Santa T (ho, yo, yo) came out of my desire to plain and simply have the viewers experience a Santa of colour; one that was extremely mischievous, fun-loving, and sported some extra booty to boot. He made his appearance annually and was rarely out of sync. Some of his early appearances have gone down in Paula's book of the funniest moments on *X-Tendamix*. She helped me fashion his dark black beard out of an old wig we had lying around from Halloween. It's now hanging on a hook in our basement (along with his prosthetic belly and bum), but I can break it out at a moment's notice. He's still good to go at the sound of a jingle bell.

The next alter ego I came up with was Selecta T. I thought I'd be able to make a special connection with my Caribbean audience with this one. But he ended up being the alter ego that I was least proud of. He was a shifty dancehall DJ and a proud Jamaican. But I was too close to the character, so he lasted only two shows (which is why he's only worthy of one or two lines in this book). Some actors will tell you that sometimes when you take on roles that are too close to your own personality it's sometimes more challenging than if you were to act like something you know very little about. I could have been an Alien T much easier than a Selecta T. You always have to give more effort and be more self-aware when you're crafting an alter ego that's so close to your own personality. When I look at the old footage of me as Selecta T, I'm like, man, I can't believe I pushed the envelope that far.

The bottom line is this. My alter egos were all fictional characters, but once I left the MuchMusic building, I wanted to be called Tony. People always expected the same kind of alter egos and cartoonish characterizations on the street — they still do — but on the street, I'm Tony. Tony who eats three square meals, works hard, and is a fan of the Toronto Raptors, just like you. Teens see me on TV and they think I'm just a little bit older than they are (as one of the band members from No Doubt exclaimed when I told him my age: "Boy, black don't crack!"). They assume I'm a young, high-profile industry type who's not about any kind of conventions at all, so sometimes I have to break it down right from the top when I speak to them. It freaks some of them out when I tell them my age. It's like they can't believe I'm not a 21-year-old hip hop junkie. Oh well, that's what you get for being six different hosts and having so many cousins drop in over a 12-year period.

MASTER T FOR PREMIER: PC CONVENTION COUP

Contrary to popular perceptions, life at Much was never all about bubblegum and Britney. There was a time when I used to occasionally host socially-relevant Much specials

Life at Much

like *2 Much 4 Much*, delving into serious and controversial topics. I'd have to say that Denise Donlon is partially responsible for thrusting me headfirst into the world of the politico, because personally I can't stand politics. Whether at the municipal, provincial, or federal level, it's all a ball of wax. Feminist scholars say that the "personal is political" and the like, but it's not for me. Having said that, I did manage to earn a whole new set of MuchMusic fans as a result of my performance covering the PC/Tory Convention in 1993.

It was the fall of 1993, and Denise Donlon, Much's newly appointed Director of Music Programming, informed some Much staffers that we were going to cover the PC convention. Unbeknownst to me or my co-workers, Denise had already selected me, in her mind, as the one to go. So one day Denise called me into her office and asked me if I was interested in covering the event. "Why do I have to go?" I asked her. "I have absolutely no interest in politics and the current candidates bore me." At this point in my life, my political views were basically this: no matter who you put in office, they can't possibly account for the whole country's needs (much less influence the life of this working-class black man). I figured that regardless of who was in power, everyone should work hard to make a life for themselves. It's not incumbent upon politicians to make our lives better; it's our own duty to improve our individual situations. And besides, as Denise was asking me to do this, I was also sitting there thinking, "We're a music video station, why are we doing this?" But the coverage of the convention was a big deal for Much. It was the first time under Donlon that the station had made a firm commitment to tackle political issues and enter the political fray head-on. So, in an unconventional style, Much sent a motley crew of hipsters and tech staff to Ottawa to cover the convention. We

Much Master T

stormed the scene with our cool outfits and even cooler hair, hoping to redefine political journalism in the process.

So I arrived in Ottawa where we were put up in a decent hotel overlooking the parliamentary buildings. On the evening prior to the first day of the convention, I was sitting in my hotel room anticipating the interview I had to conduct with one of the candidates from Edmonton nicknamed "Big Jim." So I sourced him out. I read some clippings from the *Ottawa Citizen*, *The Globe and Mail*, and *The Toronto Star* (along with all the bios on the PC candidates). Considering I had never pulled off a high-ranking politico interview before, I called Denise from my room and asked her what the hell I should be doing. I was getting antsy and nervous. Plus, I knew I was no black Peter Jennings, so I needed some direction. Denise calmly told me to just go out to the press area and "be yourself." She explained to me that Much had never really done this style of reporting before, and that for us to stand out from the talking heads at CBC or CTV we'd have to put our own MuchMusic-esque political spin on things. She implored me to ask "Big Jim" whatever I wanted, and to keep it light.

After the Donlon pep talk I was ready for the world. In the morning our crew headed down to the main grounds where the candidates had assembled. The idea was to try to get some clips from these candidates without coming off sounding too tacky. One of the first questions I asked Jim Edwards — in hip hop speak, of course — was, "When you get into 24 Sussex, how are you going to decorate the spread, man? How are you gonna make the crib look tight?" This was my feeble but funny attempt to bring the suits down to the lowest common denominator. Immediately, he started laughing from the gut and his constituents started laughing as well, so I

Life at Much

figured these interviews might not be so difficult. After all, the whole idea of me being here was to de-mythologize the political situation, to dumb down some of the political jargon and cut through the double-speak. This reportage wasn't so much about getting candidates to bore us with political platforms and promises. We wanted to bridge the gap between the suits and Canada's youth. We wanted to engage our large, youthful viewership and give them a gateway into the political world.

Certainly, Donlon should get the full credit and kudos for masterminding this whole project. Kids could obviously relate to me — not to toot my own horn (well, okay, maybe I'll toot it a little) — but I was clearly one of the more popular VJs at the station during that time. So when I covered this convention in Ottawa, it seemed to create a ripple effect. It got teens interested in becoming burgeoning politicos as well. I was to be their conduit, connecting the boardroom to the streets and giving them some idea of how, where, and by whom crucial decisions get made. The general feeling seemed to be that if Master T can talk shop with the heads of state, maybe I can too, and maybe I can get involved in the political process as well. I wanted to show kids that these political types could get downright giddy about some issues, and that they liked Weird Al Yankovic cover songs too. If we could force the candidates to contend with the concerns of the youth in the process (like unemployment and access to education) all the better. In America, the MTV kids who didn't care enough to vote were being admonished to do so by their favourite network, which became famous for its "Rock the Vote" campaign to get couch-potatoes out to the polls.

All the VJs on staff at the time were in Ottawa for this convention, and I remember that we all went to the PC Youth Cabinet

Much Master T

party. It was cool to see these future leaders of Canada hanging out and drinking booze, but I had never, in my 32 years, seen so many suits and ties. This wasn't the Tory convention — it was the Suit and Tie Symposium. Jean Charest was there and so was Kim Campbell. In fact, when we got our microphones over to Charest's camp, I suggested that his group needed a cooler, hipper theme song. The rap group Naughty by Nature was en vogue at the time, so I suggested that instead of singing their hit single, "Hip Hop Hurray," the words should be changed to "Hip Hop Charest." So my Much crew of radicals took the Naughty by Nature track and sang the interpolation. This loosened up the Charest camp tremendously. But that wasn't all that got the suits freaked out.

By the end of the day the other members of the press were on to us. They were captivated by our antics and I think they started to re-evaluate the methods they used to cover conventions like this. I think we showed them that politics didn't always have to come off looking so dull and uninteresting. My personal moment of glory came when there was a media scrum forming around Brian Mulroney, the Tory leader at the time. I'd never been in a scrum situation before, so I inched my way closer to the proceedings, taking full advantage of my size. Brian Mulroney was supposed to be coming out of the building with his wife Mila, and everyone was desperately pushing and shoving like paparazzi to get a quote from him. So after manoeuvering through a sea of security people, I got caught in a moving melee of reporters. I start screaming "Yo, Brian" to the then–Prime Minister. "Hey Brian, my man, it's Master T," I shouted out as the PM was approaching his waiting car. A stone-faced Brian Mulroney barely blinked. So after badgering him with my "Hey yo, Mulroney, yo whassup?" routine, his wife Mila looked at me and

Life at Much

said, "Hi T, my kids watch your show all the time." That was a major coup all by itself. Our version of the Royal Family had actually responded to my brand of organized confusion. By the time I got around to doing Q & A sessions with the other candidates, I had already aced this event.

But I have to say that the convention definitely made me more aware of the fact that I'm black and different. I knew I would stand out here, but I didn't think I would stand out *that* much. As far as politico reporters go, there weren't any visible minorities in sight, much less journalists with dreadlocks, baggy clothes, and nice jewellery. Honestly, covering this convention taught me that if black Canadians were to become more involved in the political process, it wouldn't be such a bad thing. So it turned out to be something of a blessing in disguise. The convention helped me get the crossover appeal that would mark my days at Much. It didn't happen immediately, but people eventually started to recognize me for my talent rather than just my colour and my urban music leanings.

When we got back home to the Much offices, I could tell we had done a good job. Denise Donlon was all kisses and hugs. After settling back into my Much vibe I called Paula from the office. Apparently, Denise had already spoken to my wife to warn her that I had aced the PC convention and that I wouldn't be able to get my slightly inflated head through the door. "Aren't you proud of him?" she asked Paula. "Very much so," Paula replied, and assured her that my good work at the convention wouldn't change me a bit. And it didn't; I took it all in stride.

The station's irreverent coverage of the 1993 federal election earned MuchMusic a Gemini award and my coverage won the hearts of hundreds of thousands of Canadians. Peter Howell, the

Much Master T

rock critic at *The Toronto Star*, referred to me as a "breath of fresh air" in this "normally droll, dull occasion." Howell said, "He's hip without being an idiot. I think he's quite talented. We should see more of him." When I left MuchMusic in September 2001, John Doyle from *The Globe and Mail* penned a column entitled "Goodbye Master T, Hello Lisa and Rod." "Now, maybe you've never really paid much attention to Master T or given the slightest thought to his segments on MuchMusic," he wrote, "but he's a man responsible for a significant moment in Canadian television. Years ago, during MuchMusic's first venture into political coverage, Master T was the man who, when he encountered Brian Mulroney on the campaign trail, said "Yo, Brian, whassup?" Mulroney ignored him and walked stiffly past, as if he'd just been asked to hug Stevie Cameron. He completely misunderstood the significance of ignoring the cheerful guy from the nation's music station and thus identified himself as a snob. The Conservatives are still recovering, as I see it. I hope Master T isn't going far from MuchMusic and I hope that some day he will be put in a TV hall of fame for effortlessly being a dude and changing the course of Canadian politics."

These days MuchMusic uses my hip hop template to get serious issues worked over by political candidates. In the last federal election, thanks to Much, we found out that Jean Chrétien favours Britney over Christina, and Stockwell Day thinks fellow former pot smoker Ross Rebagliati is a terrific young role model for Canadians. These and others from a 20-question survey were posted on muchmusic.com as part of MuchMusic's continuing attempt to encourage young people to take an active role in the democratic process. With voters aged 18 to 24 making up over 2.6 million of Canada's population, it's no wonder that Elections Canada is now

Life at Much

aggressively targeting youth by advertising on MuchMusic and YTV. Personally, I still have very little interest in politics, but the PC Convention gig in Ottawa was great for my career.

Denise Donlon
president of Sony Music Canada

The thing that sticks out in my mind was when we actually went out and did the political coverage "Vote with a Vengeance." It was the very first time we engaged in federal politics and it was a very scary period for all of us because we were energetic and ready to rock and ready to engage our viewers into the whole political process. None of us had any experience with it and we weren't sure if we could actually pull it off, but we decided to go armed with our own naïvety and ask the questions and get into the backstage, behind-the-scenes situations that our viewers would want . . . and probably never had access to. So taking T to a Conservative convention . . . he was just the polar opposite of all these guys running around being entertained with Dixieland bands. Master T was running up to a guy who was formerly the finance guy for the Conservative party and getting him to do the Butterfly [reggae dance] and scrumming Brian Mulroney, diving into the middle of a scrum with all those other political journalists, the Mike Duffys and the Peter Mansbridges and all the people who had been covering the political beat for awhile, and watching T just cut right through it, and enter with . . . "Whassup, Brian?" Nobody had ever asked a question of a politician like that before. You know, it got his attention and we made a real dent there. I couldn't be more proud of the way T walked into that arena and nailed it. And you know what, the reason he was like that was T has always had a very strong relationship with his audience, he really cares about young people.

DA DANCE MIX YEARS

What does it feel like to be the pitchman, the brander for one of Canada's best-selling albums of all time and not receive a nickel for your work? Not very good. Now, it may seem as if there's something vaguely gauche about bringing up poverty amidst all this aplenty. Yeah, I did enjoy 17 productive years at Much, blah, blah, yada, yada. But when I think about my major contribution to the whole shebang, and how I've never been properly compensated for my part, it's difficult to get excited about putting up the assortment of diamond and platinum-certified plaques that are still sitting there in my basement.

In the late '80s Quality Records was one of the top dance music labels in Canada. Headed up by Ed Lebuick, Quality was famous for being able to secure and compile the hottest dance club tracks that couldn't be purchased on CD (ones that were only available on 12" vinyl or via import). Now, Lebuick came up with an amazing idea whereby he would brand a *Dance Mix*–styled compilation to my *X-Tendamix* show. It was a good fit because the ratings were high and the music I played was diverse. The goal was to capture a share of that untapped,

Life at Much

burgeoning, underground club scene. If there's one thing you should know it's that whenever MuchMusic brands anything, from compilation CDs like *MuchDance* to *Big Shiny Tunes*, the benefit to both parties is immediate and plentiful from a fiscal standpoint. For MuchMusic it provides a means for the station to get its brand into the homes of more viewers. For the label involved, these compilations can sell platinum plus (100, 000+), which means crazy profits.

So the idea to put out a CD was facilitated by Ed LeBuick and Dave Kirkwood (head of sales at the time). It was to be a completely collaborative effort in all respects, from track selection to the content in the commercials. The tracks were chosen based on a number of variables: crossover appeal, club grooves, our ability to license the product, etc. If it sounds like this was a product of marketing considerations rather than inspiration, that's because it was. Most of the artists assembled were underground dance music artists that were making an impact on popular music at the time. People like Whigfield, Livin' Joy, Carol Medina, Bananarama, Fun Factory, N-Trance, Technotronic, Urban Cookie Collective, 2 Brothers on the Fourth Floor, and Haddaway. But all of this didn't really matter at the time because Quality was famous for pitching new products and selling household items. They had literally made their mark with infomercials.

So along with Ed and the folks at Quality, Dennis Garces and Glenn Moss put together these aggressive marketing campaigns to ensure that the compilations would do backflips off the record store shelves. Based on the success of my show, the first CD was called *X-Tendamix*. I got called into the studio to record a few punchy, in-your-face commercials to help promote this dripping wet newborn. The first commercial spot I did pro bono, even though I

Much Master T

wasn't contractually obligated to do any commercials of this nature. I received some minimal compensation for the second commercial, and at the time I was okay with it because I figured that it was all a part of promoting the music associated with my show. In the back of my mind though, I knew I was doing a flat-out endorsement, the kind of thing personalities usually get paid handsomely for. Because MuchMusic ran the commercial spots repeatedly, I started to gain a name from other audiences as the Dance Mix Dude. Still, I didn't really have anything to do with the CD. I hadn't even been consulted in selecting a single track. But the Quality Records/MuchMusic partnership was, as it turned out, a match made in heaven. In 1992, the first *Dance Mix* compilation came out and it sold upwards of 500,000 copies. The resounding success of this compilation demonstrated what can be done when typically cautious Canadian marketing departments decide to go for the moolah.

Just as we headed into the creation of the second compilation the following year, I decided that I should be entitled to some form of financial compensation for my endorsement work. At the time, there was an unspoken Much rule that said you should never bring your lawyer into negotiations like this (the fear was that it might put a quick end to your career), but I felt like I needed to circumvent this line of reasoning. When you become a pitch man for something so successful — and you're not compensated — it's easy to feel somewhat used. So I had a meeting with Much, but for some incomprehensible reason they refused to give me even a tiny percentage of the profits. The negotiations were sour, but I believed in my contribution to the project, so we finally reached a deal whereby they would pay me to do the spots. I'd go as far as to argue that the visual and sound trailers I helped devise became the signa-

Life at Much

ture selling point of the series. The trailer was linked to my voice and image for five years consecutively.

Still, Ed was the only one who seemed to understand my point of view. We both recognized the infinite potential of the CDs, so I proposed that we take the CD on the road to promote it. For our first couple of trips we hired the services of Chris Sheppard, a popular radio show host at CFNY at the time, and brought artists like Los Del Mar (who sang the Macarena song and performed the widely popular dance of the same name) along for the ride. We'd go to obscure places and clubs out in Alberta and Newfoundland. This tour taught me that Canada is not just about Toronto at all. While on the road (we toured the album across Canada) I started to notice that we're very insular on the Eastern seaboard. In Toronto we're able to access all major media, but when you go out to other city centres you see how much music means to them and how little access they have to it.

To combine the tour with my show, we brought cameras to document the party proceedings. In addition to showing clips of the tour on my show, we produced a number of specials that left no doubt about the fact that, wherever you are in Canada, people know how to party. Certainly, there was always something surreal about playing hardcore dance tracks to predominantly white audiences that weren't always acquainted with the music, but they came out to represent to the fullest. DJ Dave Campbell came with us on this tour and I relied on his uncanny ability to read crowds and get the momentum going. Dave always knew just what to start out playing in order to pull the audience into the groove before he dropped some of the more obscure stuff on them. It always felt good travelling with DC and we always had each other's back while we were on tour.

It was a successful tour. We played before 7,000 people in

Much Master T

Montreal and 14,000 in Toronto. The Energy Rush Tour (at the Edmonton Convention Centre in 1995) featured acts like Bananarama, Whigfield, Carol Medina, BKS, and Los Del Mar (all of whom were accompanied on stage by DAT rather than any sort of live band). Still, I received no compensation for my efforts, so this success was bittersweet. People would routinely come up to me and ask when my next compilation was coming out. Sometimes I really didn't know what to say. Even after all of my negotiations I was given only $1,000 to do the commercial spots. That's it. Anybody who knows me knows that I'm not a materialistic person (my mother taught me that money can buy a clock but not time, medicine but not health), but nobody likes to feel like they're being ripped off. Y'know, if they had even given me 10 cents per record I would have been making more cash than many Much staffers, so maybe that had something to do with them limiting my economic pursuits. The only person who understood where I was coming from was Ed Lebuick, and for that I will always respect him.

When the *Dance Mix* compilations reached the peak of their popularity in 1998, we sold one million copies. That's diamond certification — which is usually reserved for artists like Celine Dion, Bryan Adams, Shania Twain, and Alanis Morissette. Understandably, Paula and I were trying every which way to work it but I was trapped. There was no way I could get out of the promotions gig without compromising my job. Whether I liked it or not, I had an obligation to be the *Dance Mix* pitchman. In the eyes of the station, I owed my celebrity to Much, so they thought they had every right to be calling the shots. They had made me, they believed, so they thought they could break me if I didn't comply.

Paula, who had been writing music since 1984, came up with the

Life at Much

idea to have one of our songs included on one of these CDs, which would have made us eligible for royalties. Chris Sheppard, a Canadian artist, had contributed tracks to these compilations, so why not us? I was getting real antsy in my ongoing negotiations and finally I forced myself to come to terms with the fact that this just wasn't gonna happen. Instead, I requested that my picture be placed on the inside of these albums (or even somewhere in the liner notes). I know this probably seems like a small gesture, but having my picture on the compilation would allow me to brand the product as my own, thereby raising my career profile and leaving a mark on the public. My friend Aaron Talbot, who lives in New York and works at a record label, was the one who suggested it. And nobody seemed to object, so my picture appeared on the inside cover of the CDs for three years.

A radical shift took place at Much around this time. The Much higher-ups wanted Juliette Powell, who had recently come over from MusiquePlus, to do the promo work with me on the CDs. I was hesitant at first because the project had become my baby. But Much anxiously wanted to cross her over, fast-track her into the English-speaking market. When they started to get Juliette involved in the compilations, things began to change drastically. My show, *X-Tendamix*, slowly started to get phased out of the marketing equation. The ad spots for the compilations would no longer be exclusively branded by me or my show. And by 1996 Michael Williams and Much had parted ways. The word on the street and in the building was that he had an altercation with a record exec, was insulted by some racial undertones in the conversation, and commented to exec that he wasn't a "housenigger." This record exec was alleged to have taped the conversation and it created all of these tensions in the CHUM-City building.

My opinion was that the folks at Quality were now grooming

Much Master T

Juliette to be the pitch person for the MusiquePlus version of the CD. Powell wanted that national attention, coming from the smaller Quebec market and by 1996 — the last year I did *Dance Mix* — it went diamond. In my mind it seemed like there was this movement to eliminate some of the solid black programming like *Soul in the City* because Williams was gone.

Michael Williams
former host and producer of Rap City and Soul in the City

The struggle for urban music is [like] . . . trying to lead horses to water. I've been around the world, and the music stands up on its own, is respected, is sold, and is used the world over for the same thing everywhere. For entertainment, for cultural enlightenment. There are no boundaries to this music and this music has opened up the world to us as a people and also taken us around the world . . . The response to *Rap City* and *Soul in the City* was consistently tremendous but the labels still didn't get it. They got it later on when T came along because it was just the time. They didn't know how to exploit it to the greatest of their ability . . . If you're selling cars, that's what you do: sell cars. The politics that were involved, the politics of dancing, the politics of black music, it always sounded like they didn't want to be involved with the music, because they didn't want to be involved with the people. That's the way I kinda read it. They were just scared of the whole thing, giving it any power, giving it any juice.

Life at Much

In the early days, there was always this running battle between *Electric Circus* and *X-Tendamix*, at least as far as I was concerned. *EC* was booking the same acts as *X-Tendamix* and it seemed a bit redundant for both of us to have the same artists come on our shows. The thing was, *EC* would welcome any act — even those that didn't pertain to their format — so it was getting increasingly difficult to maintain the fine line that separated our shows. Partially as a result of this, I stopped playing so much dance music on *X-Tendamix*. I knew *EC* had that covered so I'd play everything from R&B, to hip hop, to reggae, plus some old school if I could. Plus, the times they were a-changin'. I could feel the shift away from the dance, techno, and house frenzy; more of a soul and rap-based commerciality was starting to kick up in popular music. I know that my show had an unrivalled urban-music–loving constituency and I wanted to take them further in that direction.

DA MIX: DA EARLY YEARS

There have been a lot of times in this TV game — and that's what it is, a game — where Paula and I have had to strategize, to look ahead and plan our next moves. It was like playing chess. We figured that Much higher-ups were going to try to get rid of *Soul in the City* (by this time, Michael had been gone for months). In its place, *EC* was set to become the national show that spotlighted black music. In late 1997 *X-Tendamix* went through a number of time-slot changes — and this is the industry death knell. The show got cut down to one-and-a-half hours

Much Master T

from its original three, then it became an hour-long show. My theory is that the original three-hour show had become an entity unto itself, with a distinctive look that set it apart from the rest of the shows at Much. Paula and I had to think fast before the show became obsolete. We sensed that there may be a phasing out of *X-Tendamix* and the momentum would be shifted in *EC*'s direction. In response, Paula and I decided to alter the name *X-Tendamix* to call it *Da Mix*. We put together a proposal to revamp the original formula to make it come off on more of an urban tip. I say "we" because the proposal we submitted created an associate producer position that was to be filled by Paula. Through all the support she had given me over the previous five years after hours, Paula already knew the job. She gained experience in production and even learned to edit.

I submitted the proposal to Moses and Denise Donlon. Moses' comments came back quickly and he had gone over it thoroughly. He had some positive comments, but he also gave us a few things we needed to work on. When I spoke to Denise, she hadn't yet had a chance to go over the proposal, but she was surprised that I had submitted a copy to Moses. I explained to her that I had always checked in with him any time I was about to venture into something new. In any case, we all sat down for a meeting and hashed out the idea. I came away from the meeting with the go-ahead to develop a whole new vibe on MuchMusic called *Da Mix*. We were all set to go and begin development on a two-person urban department when, as fate would have it, Paula became pregnant again (much to our delight). Due to the complications surrounding her first pregnancy, Paula was instructed to stay in bed for the majority of the pregnancy. Although Roxy was unceremoniously phased out of this new equation on account of Paula's pregnancy, I brought in DJ Dave

Life at Much

Campbell, who programmed beats for the show and could bring a new live element to the show.

Da Mix also meant that I was assigned an in-house producer. For the very first time (since I had been on the air) I was able to get some paid help to produce my shows for Much. With Siobhan Grennan, my new associate producer, things started to settle into a little routine for the first time in years. It was a welcome change and an added bonus to have her on my team, especially when we were just working out the kinks in the new format for *Da Mix*. The show had been cut down to an hour (from an hour and a half), but now it was on twice a week instead of just once. We were given new graphics for things like FNVs (Fresh New Vibes) and OBGs (Oldies But Goodies), which were concepts carried over from *X-Tendamix* and now graphically animated for *Da Mix*. A brand spanking new opening was also created by my old camera buddy Dutch. By this time, the friendly rivalry had died down between the EC camp and *Da Mix* because our mandates had become very different.

This was probably a good thing. You could almost see the shady vendors on Yonge Street printing up "Dance Music Sucks" T-shirts. Like disco before it, electronic dance music was on its way out. And so was I, as far as promoting these *Dance Mix* compilations was concerned. I had done a lot of hustling during these groundbreaking years and for me it was all about fighting for what was rightfully mine. Whether it was the *Dance Mix* situation or the *X-Tendamix* situation, I wasn't going to give up until I had exhausted every possible opportunity. Once I had done everything there was to do, then I would sit back, relax, and take stock in knowing that eventually my day will come.

Much Master T

THE REAL DEAL

I've always tried to use my role as Master T to get involved with schools and youth organizations. When you're in the public eye, you're a role model whether you like it or not. So whenever I speak to teens about my personal philosophy I tell them that the most powerful tool they can use in reaching their dreams is a solid education. Not to sound cliché or anything, but staying in school and obtaining a good education is both necessary and positive. One thing I learned early on is that no one's going to hire you if you're not qualified, so you don't make it an issue.

Outside of the MuchMusic sightlines, the most gratifying thing for me has been the opportunity to do keynote addresses for schools and youth-oriented programs. Being sought out to act as a spokesperson for a large number of important youth initiatives is the highest compliment your fans can pay you. Whenever I'd get out on the road to do these gigs I always tried to tell young people to establish a foundation and surround themselves with positive energy from people who have their best intentions in mind. Hence, total positivity (or TP) is another belief I continue to share with every public appearance I make. TP: always believe in yourself one hundred percent. When the going gets tough, you can call on your foundation to get you through the rough times.

Education is a big deal when you're from a Caribbean background, because most of our parents didn't have the opportunity to go to college or university. My mom stressed education to my brother and I constantly, and she did whatever she could to support our dreams. That's why my diploma from Mohawk College is

Life at Much

framed and displayed on my mother's family room wall. She's taped every episode of *Da Mix* too, she's so proud. Honestly, it's very difficult to climb the ranks of any corporation without a sound educational background, especially now. At one point in time you may have been able to get into companies like Much or CBC based on the nepotism thing, but nowadays if you're not studying broadcasting or communications the folks in HR won't even look at you.

I'm not even trying to give the youth one of those dull, clichéd sermons about staying in school (because, let's face it, I was no honour roll student myself), but I found my passion through school and I've pursued it. Nobody knows this, but I wanted to drop out of school on a number of occasions, especially in the eleventh grade, the year I wound up failing the majority of my courses. Instead, I focused on the two media courses I managed to pass and then I did whatever it took to carve out a career path leading to the entertainment world. Like most young kids, I didn't have a clue about what it meant to have a career back then (and the process of figuring out what interests you can be equally daunting), but I stuck with it. Like in the case of so many kids, my parents divorced early on. This meant that I didn't have the easiest time in public or high school. It must be even tougher being a teen nowadays. Fashion has created powerful peer pressure. When I was in school, wearing Levi's jeans tucked into a pair of Kodiak boots was the style. Now kids are wearing $150 Nike or Reebok shoes and driving SUVs to school.

There's a reason I've made a life-long commitment to assisting youth in the crusade against drugs, in aiding anti-racism campaigns, and in working with several other charities that assist children. Literally, I've spent hundreds of hours of my personal time volunteering at youth-oriented events. It's because I've always realized that, as one of the most recognizable television personalities in Canada's music scene, my message might resonate more with youth, than if it were coming from an out-of-touch suit and tie. I've always been of the opinion that when you speak to youth you almost have to speak their language, or your credibility disappears fast. If we as adults can slip something into the mental suggestion boxes of all the youth who perpetrate violence or are considering dropping out of school, then we've done our jobs. Another important aspect to working with youth is to give them a voice. Sometimes, top-down, adult-driven, knee-jerk responses to youth issues can potentially become a part of the problem, rather than the solution.

Through the course of my years on *Da Mix* I brought a number of youth activists as well as black teenaged achievers and entrepreneurs on board to send their messages and ply their wares in front of a national audience. I did this because I wanted to offset and remedy the mass media's potentially harmful portrayals of young people — black males in particular — with stereotyping and all. It's important when working with youth to give them a voice. Does anyone ever stop to ask them what they think about some of the situations that affect them? As one of the few black male members of the media, I did this stuff out of choice, necessity, and need. I believe in my race and my culture, and in anything I do, I always have to go home to face my community. My advice to youth who encounter racist barriers in front of them is this: strategically choose and selectively challenge the racist elements that you encounter in Canadian society.

In my opinion, one of the best things we ever did at Much to critically engage youth over socially relevant issues was when I hosted and programmed *The Real Deal* in 1993. When Denise

One of my favourite family photos. My parents had to do some serious convincing to make it happen. As you can see, my brother Basil (age 4) still wasn't amused and I had to have my ball, or no picture . . . (I was only 2).

Aunt Sis and Uncle Rupy looking cool in the '60s. I'm so grateful to them for enabling us to come to Canada.

The three lads: Basil, our bratty next-door neighbour Philip, and me posing outside the caravan my mom rented in Blackpool. Check out the legs and them glasses.

Basil and I had recently returned from the U.S. visiting my dad. My mom, looking like Foxy Brown, wanted us to take an official photo for my dad and Grandma Dawson.

We received the tour of the NBC studio in New York from our uncle Lloyd, who was the editor for *The Young and the Restless* at the time. This was a trip I'd never forget.

Posing outside the NBC studio on our first trip to the Big Apple in 1973. We were so blown away with this city, especially coming from the UK.

Studious 'T' in grade 9. This is the only picture I own of me doing homework. I had a great year. . . but let's not talk about grade 11.

I call this one my Q-Tip picture because of my cool Afro and slim frame. We won the city championship and I was the highest scorer on my team that year. I still play, but the ball seems to move a lot faster at my current age.

This was one of the Kitchener All-Star soccer teams I played for at 15 (in 1977). My coach would always put me on the field 15 minutes into the first half so I could study the opposition, and I actually played better that way.

Is that device in my hand a girl detector? Yeah, right; not even that contraption could help me with the ladies in my first year of college. It's an ambient mic and I was working at a Mohawk College basketball game.

This was my graduation picture at Mohawk College. I'd just finished working an all-night shift at my new job at Channel 47 in Toronto, and I look bagged.

My first actor's head shot ('83), photographed by my brother Basil. I've always liked this shot, but what was I thinking . . . an actor with a tooth missing?

Holding down my first job as a VTR Operator at Much back in '84. I used to crank those headsets whenever a hot video would come on.

My good pal Gord McWatters and I were both MuchMusic cameramen back in '86. A year later we started to work on the "MuchMusic Groove" epic.

Paula and I had been dating for about five months when we attended my brother's graduation from Humber College. Who'da thought we'd be together for 21 years!

One of my fave wedding photos (The Hollywood Shot). It really captured the essence of the day and we had a major party at our "Matrimonial Interlude" (a.k.a. the wedding).

The wedding party in its entirety. From left to right, my dad Karl, Paula's brother Kirk, Basil, me and my bride, Paula's older sister Donna, Paula's younger sister Allison, and Laurie, Kirk's girlfriend at the time.

I call this one my Elvis shot, taken just minutes before I tied the knot. It was a beautiful moment for me to have my father and brother as my best men.

Before Kalif was born, I used to talk to him in Paula's stomach and let him know that he was going to receive so much love. His first day home, the love was already flowing.

I love this shot; it really captures the bond between mother and child. We were out just hangin' in T.O. in the Beaches one Sunday afternoon, a few months after Kalif was born.

A rare family photo, captured at the MuchMusic Christmas party 2000. Kalif was only three-and-a-half years old.

Three generations of Youngs. Me, my moms, and Kalif soaking up Nana's Love.

Basil and I on one of our many work trips, catching some down time in front of Buckingham Palace in London, England.

On the set of *Don't Look, Cow* in Westover, Ontario. My character, T Baby, wasn't acting in this shot. Just looking at cows pee and take a dump all day in the mud... well, my face says it all.

My first trip to St. Kitts, West Indies, with MuchMusic contest winners around 1995. The volcano climb was one of the highlights. This shot was taken halfway up, but I'm proud to say I made it to the top.

Erica Ehm and I at 99 Queen, the original CHUM-City Building. This was taken in the studio around 1986. The Ehmster and I always had a special little bond.

Master T and the Super Hip Three with Mix Master Bas and our manager Dutch. From left to right, Dutch a.k.a. Gord McWatters, Richie Baby a.k.a. Richard Oulton, Low Tide a.k.a. Dave Murphy, Steve Snare a.k.a. Steve Vogt, Basil, and me. The super group of the '80s.

Entering the political arena at the PC Convention, I interviewed Jean Charest with his wife. This picture doesn't do a black man justice.

Denise Donlon, Ed the Sock, and T McGhee hookin' up for the last tree toss I participated in at Much, sometime in January 2001. Ed and McGhee actually anchored that hysterical event.

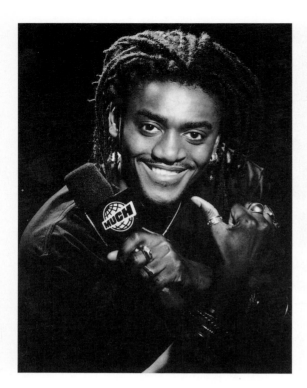

My VJ postcard. These postcards were on sale in the CHUM-City store for promotion and a new picture was taken every few years. I call this one my ode to Much, as I flex the mic.

My very first interview with the Spice Girls in 1997. This was the interview where I got a breastful of Scary Spice . . . can't you tell by my big smile? From left to right, Sporty, Posh, Scary, me, Baby, and Ginger Spice.

What an unbelievable journey to reach this moment. Sitting with a pregnant Lauryn Hill at the Goodbye Blocko, August 25, 2001. This picture says a thousand words. (Photo by Barry Roden, courtesy MuchMusic)

Mary J. Blige and I in the throes of her very first *Intimate & Interactive* at Much. We had done "In Da Round" with her before but on this night she put it all out there for the fans taking in this *I&I*. It was unforgettable. (Photo by Barry Roden, courtesy MuchMusic)

K-Ci and JoJo were two of my favourite guests to interview. Their two distinct personalities always made it a challenge to get the best out of them. This picture was taken after their second "In Da Round" at Much, where they tore it up in fine style. (Photo by Barry Roden, courtesy MuchMusic)

It was a pleasure to meet and interview Lenny Kravitz. This picture was taken in Bravo before our *Live@Much* interview. There was a mutual respect between us every time we saw each other. (Photo by Barry Roden, courtesy Tony Young)

Rick the Temp (or as I called him, Rickie Young Cat) and me at the MMVAs in 2000. Just takin' in the night and enjoying the moment with my friend and fellow VJ.

This was the final postcard picture taken of me in 2000 and is undoubtedly my all-time favourite shot. I had come a long way, baby!!

Feeling proud, positive, and strong approximately six months after I left the airwaves.
(Photo by Steven Carty, courtesy Tony Young)

Life at Much

Donlon succeeded John Martin as Much Program Director, there were different energies in the air. Now that's not to say that Martin was some detached, apolitical PD, but social relevance was at the heart of Denise's programming philosophy. So I pitched her on a show that would be designed to tap into a 12–34-year-old demographic with the intention of exploring issues like racism, sexism, and homophobia. When your PD is an unapologetic politico, it's an easy sell.

The idea for *The Real Deal* grew out of my visits to high schools throughout Canada. I found out that teens have a whole lot to say about some very pressing issues and it made me realize that they want to be listened to and respected as equals. A lot of the teens I came across felt like they didn't have a voice; their views weren't being validated. The idea for the show was also conceived of through interaction with the customers and Cribb Members at T's Cribb. I was also a part of POWER (Personal Opportunity With Educational Rewards), a stay-in-school program at the Boys and Girls Club in Scarborough, and after that I did some work for the federal government's Human Resources Development Centre (HRDC), a program for at-risk youth that was put together by activist and producer Derek Luis.

Derek and I put our heads together and *The Real Deal*, a monthly teen-styled talk show that tackled youth-related social issues, was born. It wasn't meant to be your typical audience participation show on your average TV network. It was a turning point for Much because it wasn't just a single-themed video show. We weren't necessarily gonna turn Much into the Nation's Talk Station or Talk TV, but we could now provide a window into what young Canadians were thinking. It was a 90-minute forum on issues that concern

Much Master T

teens and young adults, telecast on the last Sunday afternoon of every month. Produced by Jim Shutsa, *The Real Deal* was a hard-hitting, nationally televised program. I had worked with Shutsa on a number of productions before this (like *Snow Job* and the PC Convention coverage; we had a great working relationship because he had this ability to tap into my serious side — without losing the comedic sense). The platform for the show revolved around the fact that I would be a street reporter, and the aim was to get young people to talk to each other about pertinent issues. We combined pre-produced field interviews (I would research every given topic and conduct the interviews that would be incorporated into the show) with phone-ins and an in-studio discussion that gave youth a chance to speak out on social problems like sex, racism, teen crime, and education. The audience consisted of about 30 high school students and featured guest panelists.

I was deeply saddened when the show wrapped up because, while it was very labour-intensive, it was also morally fulfilling. With this show I was able to give something back to the community, something I believed in. If it hadn't been for the funding issues, who knows how much change we could have brought about. The most important message I have to impart after being involved in such a dynamic and influential program as *The Real Deal* is that teens must take a stand on issues that they deem critical. Never mind what adults think. For instance, if there's a racism problem at your school, what are you going to do about it? Are you going to do what we're taught to do, sit back, shake your head, and say, "That just ain't right" and move on? No. Take a stand. Use your voice.

Life at Much

SOUL TRAIN MUSIC AWARDS

The Soul Train Awards are held annually in the first quarter of the year at the Shrine Auditorium in Los Angeles, California. This annual event holds special significance for me because I get to see a high percentage of production crews that are all African-American, and most of all, I get to experience producer Don Cornelius in his element. It's definitely empowering for me from a cultural standpoint. It's a premier event that makes me feel proud to be a black man — a whole night built around honouring and celebrating black achievement. The Soul Train Awards are also noteworthy because it's where you find all the stars on the rise, celebs who have achieved a relative amount of success in the urban industry and are ready to crossover into the mainstream.

We covered this event my last five years at Much. Much staffers usually got there three days in advance of the show so they would have a chance to get acclimatized to L.A. and have an opportunity knock off some interviews. We'd also pre-tape a segment called "Much in L.A." for the regular-flow Much. While the awards are taking place, vjs rarely, if ever, get to go to rehearsal or actually partake in the show's festivities; the sets are closed. As a vj you don't even get to see the awards ceremony, you're simply relegated to handling the traffic and getting pre-show interviews on the red carpets. To see the actual show you'd have to go backstage and view it on a monitor.

When we would set up our Much booth for the awards we got there two to three hours before the event was to start and would head to the post-interview area. It's a little media area that the organ-

Much Master T

izers usually set up, with one section for photographers and another for the radio and TV press. There are cubicles constructed for the various media outlets, and you dress up your cubicle with your station's banners and logos. The associate producer accompanying me that year would be responsible for setting up our booth and decorating it. It is at these booths that we would snag the artists, during and after the festivities, to get their take on the evening's developments. In the past it was a bit of a fight to get A-list artists to come to the Much booth because even though some artists were kinda familiar with me, we didn't have as much credibility stateside as BET, The Box, or MTV. As well, this job of trying to lure celebs to our booth was primarily left up to my associate producer. She would have to be super aggressive and persuasive to pull in the artists or convince the handlers and promo staff to give us the time of day. If not, they would just walk their talent right by our booth and we would feel snubbed.

After hosting *Da Mix* for many years it became a lot easier to snag A-list entertainers that some of the big-league American media outlets struggled to get. I had developed a rapport with numerous artists and had a history of interviewing others many times over the span of their careers, not only in Toronto, but all over the world. To start off the festivities we'd go to the arrivals area with *Entertainment Tonight*, *Access Hollywood*, and BET and get into position. The Soul Train Awards were usually pre-taped; the show starts at 4 or 5 p.m., and we'd be in the arrivals area one hour prior to that. The excitement would build as limo after limo would arrive and star after star would slowly ease on down the red carpet and the press would be verbally informed as to which artist would be coming up next. Let me tell you, the outfits range from hip hop chic to haute couture to,

Life at Much

what were you thinking? All in all it's one fabulous, glamourous, soulful event. Once you've realized that the crème de la crème had worked the arrivals area, then we'd go backstage to the post-interview area and get clips of artists' reactions to wins. Backstage is all about the hustle. Not only are we there trying to do our jobs but there are also stage hands changing the sets for the show, contest winners hanging out waiting to meet various artists, celebs and their entourages doing their thing, not to mention all the press conducting interviews. This combination sometimes makes for a loud and confusing atmosphere. Other times, all we can do is hurry up and wait.

Before I even get to L.A. my threads always have to be in check. I usually line up a clothing sponsor in Toronto that deals with designer suits, 'cause straight up, the artists are judging you on your attire too. You never know if the deciding factor for a celeb to come over and converse with you came from your presentation, so I aim to look my best. I've covered a few Soul Train Music Awards with my brother Basil on camera. I've also covered it with Siobhan, Sheila Sullivan (we'd call her the Bull Dog because she was great at hunting down the celebs), and twice with Sandra Halket. One of those years with Sandra, we made it a family affair. Her husband and daughter came along as well as Paula and my son Kalif. Both kids weren't even walking at the time. It made the down time so much more enjoyable.

What also helped quite a bit when trying to secure many of the artists for interviews was if we as group were checked into the same hotel as the celebrities. One year Boyz II Men had lodgings just doors away from us. It certainly helps in snagging a big interview when you're staying at the same hotel with these stars. We get a chance to build up a little camaraderie with them, and in this industry, wearing dreadlocks certainly didn't hurt my visibility.

The guy who used to annoy me the most in the Soul Train press area was Byron Allen, the infamous host of *Kickin It with Byron Allen*. He used to always detain, no, hold hostage the celebrities, and interview them for marathon-long intervals. By the time we would approach them for a long-awaited interview, many artists would be physically and mentally drained. One year, I spotted Luther Vandross in the general press area and the next time I looked he was sucked into Byron Allen's cubicle, not to be seen again for a good 40 minutes. When Luther emerged, he was visibly disheveled and cranky. I tried waving to him, hoping that he would recognize me and come on over, but reality struck and at that point I knew he was making his getaway. Plus, remember how I was talking about keeping your look tight when you're working the Soul Train Awards? Well, every year it seemed like old Byron would always be wearing this corny black leather jacket. Go figure.

T'S VJ PRIMER

As a mature student at VJ University, I always used to pick up pointers from the old-school VJs that came before me and used them to develop my own style and delivery that branded the Master T image. With new blood coming on the block throughout my years at Much, I was privy to all the unique approaches that enabled many of my comrades to make their own personal mark on the nation. This

Life at Much

insight gave me the ability to move with the times, keep believing in myself, and share with my fellow vJs in the pursuit of entertaining and affecting the viewers (none of which would have ever be possible without the dedication and hard work of the producers and crews at Much). Here's a little insight into the Much vJs over the years from my perspective.

Christopher Ward

The vJ that I worked most closely with in my early years at Much was Christopher Ward. He was a natural ham with a trademark hyper-cool bob haircut. Thankfully, he conceived one of my personal favourite comedy sketches, the annual "Le Fromage Awards." As host Del Valour, he'd honour all of the cheesiest videos. I also admired his research skills. He would have every detail of his subjects memorized — right down to their shoe sizes and favourite cereals. You have to remember, this was pre-Internet — before the era of search engines. As far as his dress code went, it was as eclectic and interesting as his personality. Before his stint at Much, Christopher worked at *City Limits*, and when he made the transition to Much, it appeared to me that he wanted to continue all of the wacky alter egos and characters he'd created for *City Limits*. There was one funny skit we used between videos where I played the character of club owner and emcee Carlos Bassolone, and he played a character named Shecky Green, a cheesy comedian. We also did a skit called "The School of Bruce," where Chris would provide unsuspecting Much viewers with tips on how to become Bruce Springsteen. I played Clarence Clemens, saxophonist for The Boss's E Street Band. It was crazy because I worked on these slapstick pieces after my regular work shift at Much. I literally stuck around the Much building until 2 a.m. many nights, just for the love of the craft. The funny thing about Christopher is that off camera he'd become a seemingly regular dude. He had this silent intensity about him, or maybe it's just that I'm convinced that most vjs build an invisible wall around themselves as a coping mechanism to deal with life in the public eye.

J.D. Roberts

J.D. was as thorough as Christopher but he had a bit more edge, in the rock and roll sense. Many people probably don't know this, but the word around the Much building was that he was a perfectionist — you can't end up as a lead news anchor on CBS if you don't strive for perfection. As host of the headbanger-fuelled *Power Hour*, J.D. was our heavy metal dude, minus the long shaggy hair and Def Leppard tees. He wasn't nearly as outwardly wild and zany as the other old-school vjs. J.D. was the consummate professional, so it's no surprise to Much insiders why he's now one of the leading news anchors in the usa.

Michael Williams

Michael Williams was Much's "soul brother number one," before I held that coveted title. Because of the scarcity of blacks on TV in

Life at Much

Canada, there were tremendous pressures and Vince Carter–high expectations for the few that were on the air. In my estimation, it was a bit of a challenge for Michael to be representative of the black community in Canada because he was American, had grown up as a rocker, and he was not in the least bit interested in pigeonholing himself by listening to, understanding, or playing only stereotypically black music. He was way too intelligent for that. The guy was like a musicologist — he lived, ate, and breathed music, sometimes to a fault. He was, hands down, one of the most knowledgeable vjs I'd ever seen and he knew the history of all types of music.

Dan Gallagher

People think that all vjs get along, but our personalities are all so diverse that we don't all hold hands and go for drinks after our respective shows. But these rules did not apply to Dan. He was the one guy I felt most comfortable going for drinks with. He was so sociable and kind off camera that it was frightening. Despite his fun-loving demeanour, he was as serious and smart a worker as they come, and he was the one vj who didn't need the gig. His brother worked for Labatt's and they both understood other elements of the industry as far as having creative control over your ideas. For example, he conceived of the *Test Pattern* game show, and his own production company sold it to Much. Dan gave me a lot of insight into the industry and was the first vj to inform me that I should be getting paid for my *Dance Mix* promotional gigs. That's why I often confided in him and asked him for career advice. I bumped into him years after he left Much and he told

Much Master T

me how happy he was with his life because he wasn't as guarded as most vjs. My most memorable moments with Dan were our annual New Year's Eve "Time Zone" dances on Much. Short segments were pretaped and broadcast to celebrate New Year's in each different time zone in Canada. He and I would choreograph these silly little dance moves and, what can I say, Dan was as nimble on his feet as I. You know how they say that white men can't jump or dance? Not Dan. When I heard the news of his passing I was numb. Still am.

Erica Ehm

I watched Erica grow and develop more so than any other vj. I'd say that during her tenure she had a much tougher time simply because for years she was the only female vj at Much. The new school of female vjs like Jen Hollett, Hannah Sung, and Amanda Walsh have her to thank for fighting silent battles. Erica always fought the good fight, took the moral high ground on most issues, and it wasn't until her latter years when she really started to nail the larger interviews. Before there was such a thing as *Fashion Television*, Erica programmed a segment called *Fashion Notes*, and she was very big on literacy, so she'd quiz musicians on what books they read — yes, some musicians do read in between grabbing their crotches and swilling booze all day. I loved her feistiness because I like people who stand up for their rights and views, but the word on the street and in the building was that she was somewhat moody, so you had to know how to deal with her. When I was still doing camera work, I felt for her because she had to deal with way too much uncalled for bullcrap. Being a smart, female risk-

Life at Much

taker who challenges the medium is not accepted well by everyone.

Sook Yin Lee

As far as the-old school vj fraternity goes, when you look at Much, it's like a sports team. Each player has a role, and by the end of her tenure, like Sinatra, Sook Yin did things her way. I always admired her for that. Many people may not know this, but Sook Yin is a true renaissance woman. She's a former singer/songwriter from Vancouver band Bob's Your Uncle, an award-winning filmmaker, a great stage actor, and an even greater performance artist. There's a probationary period of six months to a year when Much vjs really find their own niche, and I remember that Sook Yin hit the maelstrom quickly, fast-tracking through this awkward adjustment period. She's so multi-talented that I'm actually surprised she even stuck around as long as she did, to be quite frank. When Sook Yin started getting all glossy and wearing nice make-up I was worried, like maybe this talented indie alternative chick was going too Much, but that wasn't the case at all. What's also significant to me, as far as being a media practitioner of colour, was that she was the first Asian vj on Much, and I like how she handled that role. She was seamlessly Asian and proud, and, like me, didn't really need to subscribe to the flag waving. Sook Yin's silent motto must have been: "rate me on my talent, not my race or hairdo." She was a talented vj who just happened to be Asian. I called her a week before she left Much and she shared with me her feelings on how she never really thought she'd hit the storied glass ceiling, but it happened, and she'd done all she could do at Much. When she left, it was symbolic for me because she was the last VJ from my era. She left two months after me, and it also meant that Much was losing one of their better interviewers. Sook Yin had developed a great interviewing style and an even better relationship with the alternative music community. My lasting impression of her comes from an exclusive Radiohead interview she conducted with Thom Yorke, and the interview was done at her apartment. She aced it, and left an indelible impression on all our psyches with that one.

Rick "The Temp" Campanelli

As far as the job of VJing is concerned, in my 12 years I'd say that Rick was the closest to me from a stylistic standpoint. We shared the same philosophy when it came to audience and fans, as we both are very audience-centric and fan-friendly. In the Much environs, everybody used to say that he's a shorter, whiter version of me, like a Master T without the melanin. Young up-and-coming VJs should take note of how relaxed and composed Rick is when he's interviewing big-time celebs. It's a skill not everyone learns quickly, if at all. To you, he was known as The Temp, but I called him Ricky Young Cat. Rick has a degree in physical education from Brock University, and, like me, he's a sports nut. We share the distinction of being two of the few VJs who ever played on the CHUM-City soccer team (Kim Clarke Champniss played briefly). One night when Paula and I were at Much, Rick confided in me about personal issues, and asked for career advice. I felt so honoured. Rick

Life at Much

has the ability to successfully go in many directions in the entertainment field. It's just a matter of time.

Namugenyi

Once I left, the role of filling my size 11 shoes was given to Nam. From what I've seen, she's done an admirable job under weird conditions. Her transition into becoming the new urban insider was made even more difficult for her, because she didn't have the opportunity to start fresh and develop her own show. Obviously *Da Mix* was Master T branded and affiliated, so it couldn't have been easy to host the show I conceived, developed, and hosted for six years. I'll admit, watching the transition was difficult, but I gave Namugenyi a piece of advice: know why you're here, and figure out what you can do for the urban community, because they will definitely be looking to you for something. Does she necessarily have to represent the urban, black community? Maybe, maybe not. It's purely individual. But being a black female vj at Much is not to be taken lightly.

George Stroumboulopoulos

George Stroumboulopoulos is the new quarterback of the Much vj broadcast team. He's smart, fluid, efficient, and a great celeb interviewer. By most accounts, he's one of the most competent vjs on board right now. He started at Much months before I was leaving and is handling himself well.

Ed the Sock

We, as Canadians, are so conservative, and I always envy Ed The Sock because he's the only piece of cloth in the building that can get away with articulating what most vjs feel, and he'll never get pink slipped. Whereas Ed is notorious for slamming boy bands on a daily basis, I remember making a comment about how bad New Kids on the Block were, and it caused a bit of in-house controversy. This self-proclaimed "sockarazzi" was at his zany finest when I co-hosted *Tree Toss* one January as T McGhee. For the *Tree Toss*, we'd take a whole day of programming that involved tossing the Much Christmas tree off the roof and having it blow up before falling into a dumpster in the parking lot. Ed was at his witty, brilliant best and demonstrated to me that he's no ordinary sock vj. Ed's a gutsy sock who, with Hawaiian shirt in tow, one year smartly advocated on air that Much should step up its coverage of Caribana. The Caribana coverage issue was a sore spot for me at times because I was always confused as to why we didn't provide ample coverage of the largest carnival-styled festival of its kind in North America. Ed rocks, ahem, raps.

Racial Diversity and Urban Music in Canada

INVISIBLE MINORITIES IN IDIOT BOXES

It's time to wake up and smell the coffee, Mr. Charlie and Ms. Daisy. Toronto boasts an ethnic majority of people (more than 50 per cent of our demographic is non-white). Torontonians come from 169 different countries and speak 100 different languages, and even these statistics tend to underestimate the figures. The fact that this isn't reflected on our TV airwaves is a disaster in the making. After being in Much Land for 17 years and being able to look at what's been going on at our competing Canadian news, sports, and entertainment networks, it doesn't even remotely look like modern-day Toronto, or Canada for that matter. The prime-time TV landscape desperately needs to resemble our cultural mosaic. The UN consistently ranks Canada as

Much Master T

one of the best countries in which to live, but are these spoils being enjoyed by everybody?

As an African-Canadian, you either learn to live with it and let it crush you, or you keep it moving. If there's one thing black people learned from slavery and colonialism it's resilience. I don't view race issues as something that's gonna hold me down. You have to take what you learn from it and move forward. And showing non-blacks that we are diverse and that we're way more capable of succeeding and making a mark than they might think is a big part of this. Race isn't just about black people and white people. Race is the haves versus the have-nots. The playing field for blacks and whites is so shamelessly uneven that to ignore it would be to ignore the need for justice.

Moses Znaimer

City TV was founded on the notion of diversity. In the early '70s I hired Gord Martineau and another network that was interested in his skill set were asking him to shorten his name to Martin, and that's 25 years ago. Gord's been here forever as well. And Moses Znaimer [he pronounces it out] called him to come and work here and [said] you don't have to change your name. So it's a measure of how far we've come that I'm just not a white guy and the issue of diversity has obviously become an issue of colour as opposed to ethnicity . . . of course, the factor . . . what attracted me to Tony and Basil is that they behaved like they had a right to be here. It wasn't even an issue when I gave him the mantle and he became a representative.

Racial Diversity and Urban Music in Canada

> When I hired T it was natural and appropriate and I'm pleased and proud as hell that my influence at City TV and my own experience as a Jew and an immigrant made it all clear to me. I didn't need any CRTC pressure or any studies and I didn't need all of that stuff to do the right thing. At the time I said it is the right thing and that's why we're gonna do it, and you know what it's gonna turn out to be: the smart thing, 'cause I could tell what was happening on the street. Ethnicity at its best and diversity at its best is that everybody brings their value-add into the mix and the most interesting thing is the blends, the progress, and unexpected excitement. And I'm happy to say that we're still at the forefront of a movement in broadcasting that, I believe it's fair to say, we invented.

As you might expect, racial politics plays a part in TV Land as well. Not only is there a lack of visible minorities on the air, but the people of colour in positions of authority at most networks are few and far between. Canada prides itself on multiculturalism, but where are the high-ranking African-Canadian producers and directors to be found? Real diversity just isn't there, with the exception of networks like CFMT. CHUM-City has been an industry leader when it comes to putting forth that ethnic mix, but most of the competing networks give much talk, but provide few results on diversity.

The issue of diversity in TV Land is pretty bad, but in print, where a lot of my colleagues work, it's equally unsettling. Until recently, blacks accounted for only 16 of the 2,620 professional journalists working in 41 mainstream newsrooms across Canada, according to a survey conducted by the Canadian Daily Newspaper Association

Much Master T

(now the Canadian Newspaper Association). And it's disturbing how the portrayal of our communities is so unevenly distributed. And few make any contributions to business or have noteworthy lifestyles. Considering that minorities constitute the fastest growing segment of the population (and potentially the best source of new business), you'd think media execs would be taking measures to correct these imbalances. But they're not. I wonder how many more studies I am going to have to read and memorize and pontificate about until change comes.

When I went on the air in 1990 nobody had to tell me that I was black or that race issues had to make up some part of my on-air template. Does a black VJ have a special responsibility to report on "black" issues? Well, I learned early on that when you're a VJ and you're black, your own community looks at you on a different level. Obviously urban music touches everybody, but it's important to deal with the racial aspects as well. There were Canadians across this country watching me regularly (especially in the smaller towns across Canada) and I'd be their only real contact with black people or black culture. As far as breaking down racial barriers goes, I'd like to think I made a difference in some of the more racially homogenous communities in Canada. I'd like to imagine that kids would see me and say, "Master T's black, he doesn't necessarily look like me, but he's okay." My other hope was that maybe their parents would also sit down with their kids, watch my shows, and conclude that "Hey, black people aren't bad like the media reports say." Y'know, I'm not a black man first, I'm a human being first. I care about people first and that's one of the things that has aided me in my success.

I never really dealt with racism at Much in the course of my day-to-day responsibilities. You know how social protest movement

Racial Diversity and Urban Music in Canada

advocates always say that racism in this country is very subtle. Well, it is. But not only did Moses and the Much crew not buy into the racial stereotyping crap, but they never stood in the way of me introducing a semi-political platform that had really never been seen on mainstream Canadian TV. There were many ways for me to challenge the marketplace, and I took full advantage of every one of 'em. The T-shirt line I introduced on *X-Tendamix* in 1991 was one example of just that. The T-shirts helped to project some fresh, new messages (not just the usual excerpts from Malcolm X, Martin Luther King, and Nelson Mandela). It didn't hurt that there was a whole consciousness movement happening in hip hop at the time.

When I interviewed Spike Lee and we talked about *Bamboozled*, I told him that there was a certain part of the movie that had me shifting in my chair. I had to question myself as to whether I had ever been shucking and jiving for an audience. I believe that *Bamboozled* is a must-see for any actor, VJ, or black talent who has ever wanted to get into a visual medium. Spike's a deep dude and an even tougher interview. He asked me whether a part of me was selling my soul. I replied with an emphatic "no." I simply think I touched people from the heart, which allows you to have diversity in roles. But it all didn't really matter what Spike thought. Moses Znaimer had a big hand in my attitude at the time. He was just telling me to go up there and be myself. Moses gave me such creative licence. He never came down to the set after a hardcore reggae artist like Capleton or Buju Banton would be on waxing freely about racism, black power, and fighting Babylon. I think he sensed that it's a part of my culture and my heritage; it needs to be said.

Culturally what was also interesting from the beginning of *Da Mix* in 1996 was welcoming my first associate producer, Siobhan,

Much Master T

who just happened to be Irish. Obviously, when dealing with black-oriented music, you'd hope that you'd be able to work with someone who's as passionate about the genre and as culturally connected to music as you are, but I learned that it's more important to have someone who's passionate about producing and who really knows their stuff. And Siobhan definitely did. This was also the case with my second associate producer, Sandra Halket, who took over temporarily for Siobhan when she got pregnant, and then moved into the position permanently when Siobhan moved on to another job. She stayed with me the longest and we developed a great friendship and a strong working relationship. Like Siobhan, she had solid production skills, and she just happened to be Greek. When Sandra became pregnant with her second child I was able to convince the powers-that-be to hire Petal Baptiste, who had been steadily gaining experience doing work with Sandra and I on the "In Da Round" segments, editing small pieces and also taking over the responsibilities for the "*Da Mix* Guide." She was the first black woman at Much to fight her way up to a position as associate producer. Naturally, I was very connected to Petal and it was a definite benefit to have her input on the musical side.

Now, when *X-Tendamix* and then *Da Mix* finally came along, we filled a niche that was being ignored. I entered the TV profession knowing that it's not our system, so when they gave me an inch, I took a mile. I got that inch in 1990 when I was given full reign over a TV show that was designed to showcase dance music. And I ran with it. That was the only way we were going to be able to provide for the community. I still consider myself lucky to have had a boss like Moses who was committed to diversity (much more so than most CEOs anyway). I always felt that diversity needed to be inte-

Racial Diversity and Urban Music in Canada

grated into the daily working environs of all Canadians, and it was certainly a part of life at Much. With me, Mike, Monika, Sook Yin Lee, Juliette, Byron, Namugenyi, and Nadine all holding down the fort, it was clear that diversity was possible. Thankfully, through the work of organizations like the Canadian Association of Black Journalists (CABJ), a network of communications professionals committed to ensuring that the rich and varied mosaic of Canada is reflected in the media, this diversity continued to grow throughout my time on the air. Sadly, imbalances are still prevalent in many other areas of TV production. In Canada, people of colour obviously don't have the same type of representation on TV. Not on the Home and Garden Network, not on CMN. So we have to be able to put our own stories on the air.

At Much there was a lot of responsibility placed on me as the only visible black broadcaster. This has to change. No single black media practitioner can be a spokesperson for the whole race or community. Blacks in Canada are not one homogenous, undifferentiated glob of humanity; we do not all share the same value systems, ideals, and interests. The only way things are going to change, I believe, is if we take ownership of media outlets ourselves. Much like Denham Jolly and Flow 93.5 FM, my dream is to set up my own TV station, so blacks and people of colour can have some control and put some positive images out there for others to appreciate.

My determination to see the media landscape reshaped comes from my parents' experience. As bad as things have been for me at times, my mother and father had to deal with much more blatant forms of racism. Our windows used to be cracked by racists in Leeds who didn't want us in their neighbourhood. I'd be walking down the street and have people yelling "nigger" from car windows; skinheads would chase us through the streets. It was in-your-face. In Toronto, it's not as in-your-face, but racism is far from dead. When Dwight Drummond, a close friend and an on-air colleague, can become a victim of racial profiling (in a widely discussed racial incident) it serves as a wake-up call to us all. And when we're given the opportunity to de-mythologize the moronic racial stereotypes that lead to such incidents, we have to be teachers. No, I'm not a criminal. Rastafarians are not vagrant dope heads. Yes, I listen to other genres of music apart from rap. Nuff said.

HAPPY TO BE A NAPPY VJ

What was it like being the only dreadhead on national TV for the past 12 years? Well, let's just say that when your company's Big Cheese, Moses Znaimer, implores you to "shake your locks freely" and to push forth your agenda, it makes the decision to not look like every other black suit and tie on TV that much easier. Certainly, for my locks-wearing brethren, it must also be inspiring to see someone in my position sporting dreads freely, with impunity, and pushing the envelope of hair. But either way, I've never worn locks for anyone's benefit except that of me and my wife. I can honestly say that I'm happy to nappy.

I literally haven't combed my hair since 1989. Since I was in high school I've always wanted to have these curly knobs sitting atop my head. I first tried to lock my hair in grade 10 with mixed results. Without getting into too many details, let's just say that I had a bad hair year. It just didn't pan out. In the growing or "baby dread" stages I used to wear a hat, which was a pretty stylish way to cover up the uncomfortable, in-between stages. When I eventually took my hat off around Christmas 1979 (this was after three months of growing), my mother nearly had a heart attack. "Don't bring that in this house," she said. "Cut it off, I don't want any Rastas under my roof." She was not having it. But when you move out of your mom's house you can pretty much do whatever you please. So when I moved to Toronto in 1984, I was free to contemplate whether or not I wanted to start growing dreads again, and I didn't for about four years.

Before I had the full-fledged locks I have today, I sported a box cut. The sides of my head were shaved clean and the top had a tuft of Afro shaped in an inch-and-a-half box, with edges and all. It was really extreme, perfect for the times. This cut was made famous by Grace Jones and I sported it for a couple years. It was very popular in hip hop culture in the late '80s and early '90s. A more two-dimensional version was worn by Kid from the hip hop group Kid 'n Play. I definitely was ahead of my time. When I got tired of that look I decided to start growing locks, but to be different I still kept the sides shaved and just let the dreads spring up from the top of my head. For many years while I was on-air I continued to shave the sides and the back off for a Mohawk effect. I got some major flack from black communities at home and abroad. In the early '90s I went to Jamaica after my marriage and we stayed in a resort in Montego Bay for our honeymoon. I was getting pure grief for not growing out my full head of locks. It was virtually sacrilegious, according to Rasta religious tenets, to manipulate the hair that The Most High (read: God) had given you. So haircuts, buzz cuts of any

Much Master T

other kind are forbidden. The locals called me "Half-Lion" and "Rent a Dread." Holy cow, did I not realize the extent to which this "fashion dread" labelling process would go on. They would constantly badger me with the "Half-Lion" moniker (Rastas call themselves the Lion of Judah based on Old Testament scriptures). Some would insist that I must be eating pork because of my hair, which is the ultimate no-no according to the Book (like Nazarites in the Bible, Rastas are into pure and Holy living and they can't eat pork). The vendors in Jamaica, whom we didn't buy anything from, were especially rude.

Anyway, thank goodness the half-dread confrontations played out within months. Man, that hairstyle should be put in a time capsule. For local black audiences it was just as much an issue as it was "back home." One day I decided I was gonna grow my locks all over my head and that was a tough stage. Certain elements of the black community in Toronto were just as incensed about me wearing locks on top of my head, as they were on the sides and the back. It took the black community a long while to adjust to me. Sometimes I'd question myself, but it obviously wasn't a conscious decision to wear dreadlocks for TV shock value purposes.

Wearing locks in 2002 is way more acceptable than it ever was before. On Queen Street West you can find many non-black men and women sporting the 'do. Has the hairstyle been co-opted into the mainstream and made to look like some quasi-radical stuff in a consumer friendly package? Maybe. But back in the '80s when I started growing my locks you were definitely making a statement, socially, politically, and otherwise. When I first started growing in 1988, locks were singularly associated with being a dope head, a Rasta, or in some trouble otherwise. The ignorance at the time was deafening. Despite the fact that real practising Rastas are the most peace-loving, spiritual beings on the planet (next to the Hare Krishna and the Gandhi-ites), films like Steven Seagal's *Marked for Death* still managed to connect them to crime and all that has gone wrong with Western civilization.

At the end of the day, the real reason I sport this 'do is three-fold. It's something passed down from the Masai tribes in Kenya, via Jamaica, to Toronto. Locks have roots that go back as far as the fifth century in Africa when Ethiopian Coptic Church priests let their hair grow. There's a sense of cultural pride for me in sporting my locks. For the record, I'm not a Rasta, but my hair signifies some kind of unspoken black consciousness. In the Western world, the wearing of locks was born out of the Black Power movement of the late '60s and early '70s. It made a serious cultural dent when acts like Bob Marley and reggae music entered the mainstream. Some of the principles and belief systems that the Rasta community engage in definitely play out in my life. I was not born in Jamaica, but my first visit with my wife in 1989 made me want to develop a closer tie to my culture. Buried beneath the misinformation and false mythologies about the hairstyle is the fact that not many people know that other ethnic communities sport this 'do too. East Indians called sadhus (nomadic Hindu holy men and women) also wear their hair in this "jatta" style, so it's not exclusive to black people of African descent.

Y'know, one time I was driving and my hair was near my neck in length and it was blowing in the wind and it was the most incredible sensation. There's a certain part of you that strives to be at one with your hair. You have to nurture, groom, and endure the natural growth process. It teaches you patience (it took me 10 years to get 12

Much Master T

inches in length). Dreadlocks are a time commitment. By baldhead standards, my locks routine is grueling, to say the least. I wash my hair two to three times a week and my hairstylist/wife/personal advisor twists my locks to tighten them up once a month. I'm in capable hands with Paula; her locks are past her bum now.

I'd have to say that when real Rastas come around, they can see what I'm doing. The respect is there. At Much I was like the unofficial roots reggae spokesperson, so it was all good. I remember I used to get nervous when I'd interview real Rastas like Burning Spear. He has this song called "Rasta Business" where he talks about fake dreads, commercial dreads. Thankfully, he sensed my genuine respect for the hairstyle and the culture. These days, commercial dreads are so popular that you have meetings like the California Dreadlock Conference held annually in Sacramento, where bohos and beatniks converge. Even in Toronto there are dozens of lockticians and specialty salons springing up every half hour. As far away as Tokyo, the Japanese pay thousands of yen to have their straight manes molded into the hairstyle. I hope these people take the time out to seek the meaning behind this hairstyle. I'd like to believe that my trademark dreadlocks — something rarely seen on Canadian television — helped to endear me to black Canadians, rather than annoy them. I realize that many are called to lock, but the chosen are few, so locks ain't for everybody. Should everyone who locks their hair adhere to the tenets of Rastafarianism? That's up to the individual to determine. For the record, I have no intention of cutting my locks anytime soon. The hair has become a part of the T package. It's a part of my being. It gives me a regal presence, kind of like a lion and his mane. The locks can only keep growing and getting gray. And as they change colour, I feel it will only make me look more distinguished.

Racial Diversity and Urban Music in Canada

ICED OUT

At one time I used to wear about 14 silver bangles on my wrist. Since leaving MuchMusic I've gone down to four. Over the years my regular Much viewers used to jokingly refer to me as the Black Liberace. But straight up, the only cross-dressing you'll see from me is when I mix the French and Caesar for my salad. Heck, most rappers these days (think P. Diddy or Master P) wear platinum-plated necklaces, rings, teeth, and chest pendants, so my little bit of accessorizing is not that big a deal.

My silver fetish comes less from wanting to flaunt my wealth (like some rappers do these days) and more from the fact that I'm a product of the disco, house, and new wave music eras. Whatever the offspring of punk, funk, and everything in between looks like, that would be me. My silver fixation was sparked when I was still in my youth, even before I came to Canada. There was this mechanic that I used to work for part-time in England. He gave me a two-pound bonus when I left the job for Canada. I took my two pounds (which was considered big money back then) and bought this horoscope silver ring that I'd wear all the time. I'd wear it to school, soccer practice, and to sleep. I'm not exactly sure what happened to that ring, but it remains a good childhood memory for me.

So way before I went on the air in 1990, I had stepped up my search for the perfect plated accessory. In the early '80s, just shortly after Paula and I left college, I started to collect copper jewelry. I had heard that copper bangles were a big thing in Jamaica at the time, so I picked up on the trend. Most conscious Jamaican expatriates carry this emotional connection to their spiritual homeland, Africa, and I've always had this natural affinity for the Masai tribespeople in

Much Master T

West Africa. The Masai were a warrior tribe from West Africa who had been transplanted to Jamaica during the slavery era. I read that they were feared and treated like royalty. Maybe it was the Masai influence, or maybe Paula and I were just two radically alternative folks who took the Queen Street scene too seriously. For whatever reason, we loved to have things dangling all over the place.

When I walked around the Much building with my jewellery on I felt like an African king. I'd wear wooden beads and a hand-carved Masai warrior around my neck. As far as my wrists were concerned, I used to wear 13 copper bracelets on one hand and 13 silver ones on the opposite hand at one time. I'd have to say I got a mixed reaction from viewers, fellow staffers, and the fashionistas on Queen West. Women tended to love 'em, but the fellas didn't seem to know what to think. Others used to rib me, implying that I had a Mr. T starter kit, or that I was a poor man's Mr. T. In my mind I was simply mirroring the Sammy Davis Jr. approach to wearing rings: the more the better.

One day in 1996 I decided to pare my collection down. Why did I reduce the glitter at that time, despite wearing them on-air for five years? Straight up, there was no conspiracy theory to report on, the jewellery had simply become too heavy. Every time I'd pick up a hand-held microphone to interview anyone from Mariah Carey to TLC, it felt like I was picking up a five-pound barbell. Plus I had carelessly misplaced or lost some of the VIP pieces of my collection, so I figured it was time for a change. My jewellery issues came to a head when Paula and I were struggling to determine what type of rings we'd like to exchange for our wedding vows. Picking out my wedding band was a real contentious issue because I've never liked gold, and gold is the modern-day wedding cliché. Y'know, gold rings, black ties, and white dresses. At the end of the day, I ended up getting a combination of white gold and gold for my wedding band. I had to have a little tradition thrown in there.

Sure, maybe the sheer number and styles of bangles I wore on my wrist hosting *Da Mix* or doing my regular Much shift was excessive, but more than anything it was an attempt to stand out and develop a current connection with some historical ideas where adornment was concerned. Believe you me, when I look back at my old tapes I'm amused that I could carry it off for so long and I still love the funky look of them. And as I write this, the downsizing of my jewellery collection continues at much the same pace as the downsizing of services in the Megacity. I'm now down to three rings and six bracelets on my right hand, all of which are made of brass and copper. And on my left hand, I have three silver bracelets, a watch, two silver thumb rings, and my wedding band. The biggest complaint I get about my jewellery excess is from Paula. You see, I never take them off (not even to go to bed) and she's always waking up with a couple of them tangled up in her hair.

URBAN MUSIC IN CANADA

The Canadian urban music scene is alive and thriving as I write this. I have an overwhelming sense of pride in knowing that I was instrumental in nurturing the Canadian urban flavour that is just now becoming more sought-after by American artists and labels. Want proof? Well, in just the last three years a good number of local notables who

Much Master T

appeared on *Da Mix* regularly have all signed deals with major U.S. labels. Kardinal Offishall signed to MCA, Saukrates was picked up by Def Jam, K-os is with Capitol/EMI, Glenn Lewis with Epic, and Jully Black with MCA. Watching these twentysomething cats take the business of the industry into their own hands and plot out their own paths to success has been inspirational. It's been a pleasure to watch and support them over the last decade. I've always felt it's important for the young, burgeoning acts to pay homage to the artists who helped pave the way for them to be able to "bling, bling." For all the revisionist hip hop historians out there who think Choclair was the first local act to be get accolades in the U.S., guess again. Toronto-based female emcee Michie Mee was being courted by American recording labels in the late '80s, way before there was a fully developed urban music scene in Canada.

In the mid- to late-'80s, Ron Nelson, Toronto's most renowned radio programmer, hosted a show on Ryerson's campus community station called *Fantastic Voyage*. Ron split much of his time between producing and promoting and he would routinely import acts from the U.S. to perform. It was during one of his U.S.A. versus Canada rap battles that a local star was born. Ever since, the stories of Michie Mee's skills on the mic have become urban folklore. By 1989, Michie (whom some of you might remember as the star of CBC's *Drop the Beat*) had already earned a rep as one of the fiercest emcees in Toronto. She was managed by Ivan Berry, a young, hip music exec with a keen eye for talent. Ivan and Michie were crafting up a new sound that would fuse their Caribbean roots with American attitudes, beats, and rhymes. Michie had paid her dues for a few years as a talented underground emcee, under the early tutelage of K-Force (K4CE). By the time she was put on the big stage at Ron

Racial Diversity and Urban Music in Canada

Nelson's cross-border battle, she had out-rapped America's reigning queen of hip hop, Roxanne Shante, in one of Canadian hip hop's most defining moments.

Michie Mee
Canada's first lady of rap; actor

One day I decided that I was going to New York to make it big. My family lived in the Bronx at the time and I went down and met Scott La Rock, Jazzy Jeff, and a whole bunch of Zulu Nation members. They would say, "Hey, there's that girl that talks funny from Canada." Scott La Rock met me at Latin Quarters and asked me to be a girl in one of BDP's video. And I asked them, "Why don't you come down to Canada? We have this guy Ron Nelson who throws great parties." So I went back to Canada and started running off my mouth to Ron and Ivan. I had the hook-up. Ivan started talking to Nat Robinson who had already heard about me. I had recorded a track with KRS-One called "Elements of Style" that got a good buzz and I was on my way, or so I thought. I went back to New York and recorded for K-swift, who was a producer for MC Lyte. Lyte and I got together and recorded a track. Unfortunately, in between that, Scott La Rock got killed. Scott was my Puff Daddy. My career would have gone in a whole 'nother direction had he been here. That was the first deflation of my career. I was at Flemingdon Park playing basketball and I heard the news that Scott had been murdered. From that I leaned more into the First Priority camp than BDP. I released a record with MC Lyte; they had a compilation distributed in the U.S. We got a

Much Master T

budget from First Priority and took MuchMusic cameras and staff to Jamaica. Conceptually, we wanted to do an album that was one half reggae, one half hip hop. In Toronto, every rapper was mimicking American sounds, so we wanted to move away from that and simply try to prove to the mainstream that hip hop wasn't disco, that it wasn't a fad.

When the organizers of the Harry Jerome Award for Lifetime Achievement in Music (one of the most prestigious awards the black community has to offer) asked me to present the award to Maestro in 2002, my whole life in the urban spotlight flashed in front of me. As I made my introductory speech, I reflected on the extent of the contribution Maestro has made to the Canadian music industry. I wanted everyone to know that he was the first rapper to go gold and platinum in Canada, to receive a Juno award in the rap category, and first Canadian rapper to receive national radio airplay. His influence in opening the doors for Choclair, Kardinal Offishall, the Rascalz, Swollen Members, Saukrates, Motion, Ghetto Concept, and many other burgeoning rappers is undeniable. When I closed out my speech by announcing how proud I was that Maestro is a part of the Canadian black music industry, I realized we'd come a long way, baby. When I first went on the air in 1990, the urban music industry was small and insular. It only consisted of a few key people trying to push the music and culture forward. Maestro and Farley Flex (Maestro's former manager who's now an employee at Flow 93.5 FM) were two of the first instrumental players. I watched Farley fight and struggle to get radio acceptance for his artist, and I watched him

Racial Diversity and Urban Music in Canada

dealing with a community that wasn't exactly sure where he was trying to take the music. Meanwhile, Maestro was continually plugging away, demanding respect for the music.

Maestro *rapper* Many people may not know this but my very first real video was not for "Let Your Backbone Slide," but for a song called "I'm Showing You." I put that out independently and spent $5000 on the video, which was shot on 8 mm film. I didn't know about Videofact back then and we didn't know better at the time. All I knew was national exposure back then, so if I get my video out, that's national exposure. If you're gonna be seen across Canada, then something big was gonna happen. When I got my record deal from performing in the Much building on EC in front of Stevie B and the LMR record execs (who were American), it signaled to me that MuchMusic is gonna be an important centrepiece for black music to be exposed, and whoever worked black music there as a VJ or host was gonna have a large say in some of my career strategies. When Much was making the transition from Michael Williams to Master T being the VJ for *X-Tendamix*, it's like T was at the time just coming to the forefront of being a black face on MuchMusic. He saw the responsibility, the obstacles, challenges amongst not only white people, but his own people, because people were trying to challenge his role at Much out of the frustrations of a neglected music scene. When you have a station like MTV and BET or shows like *Rap City* on MuchMusic, the questions we have to ask ourselves is

Much Master T

> how are we being represented? Who's the person here who's representing us through music? Early on, Master T knew there were a lot of challenges and negative stigmas that Michael Williams might have had to deal with, that had to be overcome.

At the same time, Devon, a popular rapper from the late '80s who creatively combined the dub poetry tradition with hip hop and reggae, burst on the scene with his politically charged song, "Mr. Metro." Released in 1989, the track focused on the issues of police brutality in the black community. Devon ingeniously fused his strong lyrical content with funk, soul, R&B, reggae, and ska influences. To put out such a controversial piece of work at the time was definitely risky. It blazed the trail for tastemakers like Ivan Berry and his Beat Factory record label. As African-Canadians, The Dream Warriors captured the black community's true diasporic yearnings on their debut album *And Now the Legacy Begins*. Sure, the album sold over 800,000 copies globally, but that's not even the most important fact here. When emcees King Lou and Capital Q took their fusion of reggae, calypso, and jazz and introduced such hits as "My Definition" and "Ludi" to the world stage, it placed Canada on the map as far as musical diversity goes. Canada was no longer just the country that produced Bryan Adams or Celine Dion. Unfortunately, it took Canada a long time to accept, and finally embrace, what the Dream Warriors had done. Domestic record labels didn't really start to embrace black music until the early '90s when American rappers like Ice-T, Coolio, Public Enemy, N.W.A, and the Wu-Tang Clan starting selling gold records without any commercial airplay in Canada. Once again, we were playing follow the leader with the Americans.

Racial Diversity and Urban Music in Canada

Michael Williams

Today, everyone wants to be "urban" and involved and interested because it's become a multi-million dollar industry in Canada. However, unless acts take their routine south of the border they get stuck in that Maestro zone. Black artists have taken on the punk ethic, the Canadian ethic of "we've gotta get the hell out of here." And we have to DIY. Saukrates, Tha Circle, [are] self-contained units with their own management. And you can see why. As far as labels go: one, they don't have the interest; two, there isn't the knowledge of the songs or production; three, when people are interested they get rebuffed constantly, as I was rebuffed for several years, and then I had to make things happen on my own. There was never any problem with the music, the problem was the people who were supposed to be bringing the music to you. Our music has been tried and tested, from Coltrane to Snoop, y'know. And the fact that people always tried to just sell black music to black people, "we don't have the population" excuses, these are all lies. One, you don't need the population because you're selling the music to whoever's out there. Hello, wake up. The black dilemma is the same as the Canadian dilemma. So I guess we should call it the Black-Canadian dilemma. What Glenn Lewis experienced is the same thing that Joni Mitchell and Neil Young went through. They're travelling a well-worn path to success. If you talk to most people I think they'll say, "You have to get out of Dodge" when you're contemplating a career. Even in small town America, you gotta get to New York, you gotta get to L.A.

Much Master T

Canadian major labels were asleep at the wheel from the mid '80s on into the early '90s. That made it all the more critical that I spin some videos from groundbreaking urban Canadian acts. I played videos by: Billy Newton Davis, Lorraine Scott, Lisa Lougheed, John James, Rupert Gayle, Temperance, Oval Emotion, Charlene Smith, Messenjah, Sattelites, Snow, Sunforce, Clifton Joseph, Donna Makeda, Michie Mee, Thrust, Ubad, Too Bad To Be True, HDV, The Rascalz, Kish, MC J & Cool G, B. Kool, Top Secret, Split Personality, Citizen Kane, Motion, Ghetto Concept, Choclair, K-os, Saukrates, Simone Denny, Glenn Lewis, Wade O. Brown, Carlos Morgan, Jully Black, Rumble and Strong, and Main Source, to name a few. In any discussion of black music in Canada, I think it's important to applaud the fact that the majority of these artists released their music independently through small indie labels and distributors. This was another big reason I rallied to give these artists some shine (in addition to the mandatory Canadian content quotas imposed by the CRTC). In cases where an artist didn't have a video out (but I felt their music was definitely up to par with the major world performers) I would invite them to perform on the show. By the time *X-Tendamix* made the switch to *Da Mix* (there were now a good half-dozen young, erstwhile urban record execs in the Canadian industry), the quality of the videos being submitted had stepped up quite a bit. Early video pioneers like Joel Goldberg (he directed Maestro's "Let Your Backbone Slide"), George Vale, and Rage documented this exploding youth culture with their wide lenses, and Allison Duke, David Cropper, Kilolo, and Little X — one of the most influential video directors in the world today — got to showcase their skills with the aid of VideoFact. The end result was that the amount of Canadian urban videos added to MuchMusic's regular rotation almost doubled.

Racial Diversity and Urban Music in Canada

By the mid '90s it had become mandatory for all urban artists to have an above-average CD, some form of distribution, and a decent video to even be considered for airplay on Much. These three key elements were certainly necessary if an artist was going to be invited to come on *Da Mix* to promote and perform. Gradually, the major labels in Canada started to understand how potent my show was in promoting this hip hop culture that permeated everything from McDonald's commercials to MuchMusic ad campaigns. Finally, they started to open their doors to young blacks in the field. I can't begin to tell you how happy I was to see a new generation of young, progressive record executives emerging on the scene at this time. When John Bronski, a local community radio host at CIUT and mastermind of the first all-Canadian rap compilation "Cold Front" on Attic Records, got a gig at Sony, I knew the doors were beginning to open. Then years later, Craig Mannix, a diehard hip hopper who had interned on *Rap City*, got a gig as an urban music consultant at BMG. This was one of the first real jolts of energy the stagnating music business needed. And it was immediately followed by the success of transplanted Vancouverite Sol Guy, who originally started out managing the Rascalz before eventually moving over to BMG in 1994. Then you had this hipster white dude named Russ Hergert at Virgin Records who was ultimately responsible for getting Choclair signed to a deal. I was also becoming a mentor to some of these guys because we were all working within an industry that was more than a little slow to catch on. Without this influx of young execs, I don't think the urban music scene would be as large as it is today.

There was also this non-musical urban cultural activity that accompanied the influx of these new, young energies. Cats like Sol Guy introduced some entirely new guerilla marketing concepts to the

Much Master T

Canadian arena (in particular the idea of having street teams posterize strategic areas in Toronto with promo materials). The industry would eventually expand to include urban fashion wear. Adrian Atcheson and Robert Osbourne, co-proprietors of 2 Black Guys clothiers at Bathurst and Bloor, were the catalysts of this fashion movement. Their politically-charged message-oriented T-shirts made everybody from celebs to the hip hop uninitiated take notice. How many black Canadian designers do you know who've had recording artists like Mary J. Blige, or filmmakers like Quentin Tarantino and Spike Lee endorse their wares? Other urban clothiers jumped on the bandwagon shortly thereafter. The 100 Miles crew, led by Gary Adamson, opened up a shop on Yonge Street and his hip, happy urban apparel did somersaults off the clothing store shelves. The collective who own the Nappy's Barber Shop chain also put out their own line, which sold well at retail stores. And of course, Paula and I eventually ventured into this line of business, opening up T's Cribb in Cabbagetown, a store dedicated to carrying all of these Canadian T-shirt lines.

Fast forward to the present. Lounge Urban Separates, a company run by Daymon Green and Desmond "Dread" Hill, has been even more successful in bringing the latest Roc-a-wear, Echo, and Enyce clothing lines to a hungry urban market. The Lounge crew has also branched to create their own Lounge brand line and communications company. Working alongside mainstream companies like Reebok, they join other noteworthy hip hop stylists in outfitting local artists and personalities including myself. From the mid '90s, right up until the day I left MuchMusic, I was exclusively styled by Lounge, so they're to be commended for making me look über-cool. Sadly, once major retail chains started carrying these clothing lines, it effectively killed the cool factor that had drawn me to the baggy, hip hop styles in the first place.

Racial Diversity and Urban Music in Canada

In the world of magazine publishing, a whole slew of urban-themed magazines came out of the woodwork at the same time to document the presence of urban culture in our everyday lives. *Word* magazine was one of the most all-encompassing in that it not only put a mirror to hip hop lifestyles, but it also reflected the broader urban culture in its treatment of film, art, and literature. I would also regularly scan other mags like *Mic Check, Peace, Pound, Klublife*, DOJO, *Vice*, and others.

By the late '90s, even Canadian DJs were starting to ink recording contracts with the major labels. But in typically Canadian fashion, these opportunities came at the expense of urban music and culture. In late 1999 the Canadian Recording Industry Association (CRIA) and the Organized Crime Intelligence Unit of the Toronto police force teamed up to raid a couple of mom and pop record stores on Yonge Street (Play De Record and Traxx). They came away with over 6000 so-called "unauthorized" mixtapes. The raid sent a strong message to the Canadian hip hop constituency. Why would they have wanted to shut down the livelihood of mixtape DJs, many of whom actually work with the major labels themselves to promote new records? Hard-line approaches like this aren't even upheld in the U.S., the birthplace of hip hop. Not only were the store owners charged, but any DJ whose contact info appeared on the seized product was charged as well. Thankfully, Babyblue Sound Crew, one of the groups implicated, were able to parlay CRIA's myopia into a recording contract with Universal Records and their debut, *Private Party Collector's Edition Volume One* went on to sell near platinum. Likewise, Virgin Records scooped up the services of another popular mixtape DJ. Mastermind and his debut, *Volume 50: Street Legal* was critically acclaimed for smartly delivering a number of collabos

Much Master T

between American and Canadian rappers.

Because of this flurry of hip hop–related activity on the streets, I became even more aware of what my role in Da Mix (pun fully intended) would be. Because I had travelled extensively across Canada, I came to realize that my show was not only a video program. It would also act as a makeshift urban radio source that could be used to tap in to the hottest dance tracks that weren't accessible on regular commercial radio. Regardless of the inherent fear of most Canadian radio programmers (who were and are still clutching on to the old ideas of what constitutes a good "pop" song in Canada), *Billboard* charts were clearly being dominated by urban music artists. By 1998, hip hop had become the pop music of the day. And it still is.

Right from my early days at Much, *X-Tendamix* (and then *Da Mix*) was the only clear-cut national outlet for urban music. As a result, I wanted to push not only the best international chart toppers, but to expose Canadian artists to the nation as well. I wanted my viewing audience to be proud of the fact that so many of the hot dance, hip hop, R&B, and reggae videos they often requested were homegrown. Still, promoting native talent didn't always come without its share of problems. My biggest pet peeve, after years of programming *Da Mix*, was that I'd receive numerous calls from some very persistent artists (who shall remain nameless) who desperately wanted to come on *Da Mix* to promote their music. The problem was: most of them would barely have completed even one of the three requirements necessary to have the *MuchMusic* machine aid them in the promotional process. I can't even tell you the number of times I got calls saying, "Master T, I have a great video, can I come on your show?" from artists who didn't have a CD in stores to complement the video airplay, much less a distribution

Racial Diversity and Urban Music in Canada

channel. As a producer of a show, you constantly have to be aware of this Catch-22 situation. If I play Joe Rapper's video and it stirs up some attention and starts charting, where does that leave the potential consumer who can't buy Joe's non-existent CD? Certainly the video won't be out for sale any time soon. To me, it seemed like a simple equation. But I still managed to spend countless hours on the phone with dozens of artists who just didn't seem to understand how to make the Much machine work for them.

In my years at Much I noticed that Canadian artists, urban or otherwise, have this fixation on trying to impress our cultural imperialist neighbours to the south. Sure, knocking on America's door is fine, but my dream was always to see Canadian artists be able to sell gold and platinum in their own backyards. When the Rascalz recorded the all-star emcee posse cut "Northern Touch"(which was the brainchild of Craig Mannix, Sol Guy, and Mike Z), they labeled themselves the Northern Touch All-Stars, it pushed the Rascalz' *Cash Crop* LP to gold. There couldn't have been a prouder moment for me. It meant that Canadian teens were not only requesting Jay-Z and Nas all day, but that some consideration was being given to artists that they may have grown up with or seen casually strolling along Queen West.

Red One — The Rascalz

When we recorded the song, the country took it on as their own. It was the national urban anthem. Choclair, Thrust, Kardinal got exposure deals from that. Originally we recorded the song for a DJ in Vancouver named J-Swing and Kemo. So me, Misfit, Checkmate, and Concise had rapped on it for the mixtape. Big C and Sol Guy heard it and between the two of them they were

Much Master T

thinking we should get the rest of the emcees, some Toronto emcees. So we sent the track to Toronto. They (Choclair, Thrust, Kardinal) recorded on it. Kardinal came up with the hook which I think was the killer for that song, which really made it what it was. When we put it out, the DJs were going crazy and the reaction from everybody was that we have to make a video for this. It was crazy because this song wasn't on the original *Cash Crop* LP. We stuck it on after our album was at 35–40,000 sold, and it pushed us past gold. Little X shot the video and we'd known X since we first came out to Toronto and he was a graffiti artist in this group, and we had a graffiti artist (Dedo's in our group). Due to the similarities we just bonded the two crews. He really did favours for us to get that video to where it was. At the time he was shooting big budget videos for Hype Williams. "Northern Touch" raised the bar. It got most of the artists in there deals. It was the only local product to be playing on BET every time.

The changes in the Canadian scene were rapid-fire from there on in. Choclair was signed to a deal quickly afterward by Virgin Records, who recognized early in the game that hip hop was the pop music of our day. When I was granted the opportunity to present Choclair with his gold record for *Ice Cold* on *Da Mix*'s "In Da Round" special, it was like my dreams were starting to materialize right in front of my eyes.

Choclair
Juno award–winning rapper

Racial Diversity and Urban Music in Canada

When I went to MuchMusic on December 9th to do my first live TV appearance, it was weird because after the show, I saw the president of Virgin Records hanging around. And there were other Virgin staffers in the building. And it didn't click in yet. I asked one of my Virgin reps why they were at MuchMusic waiting for Master T to come by and they lied to me and said they came by to lend me support because I was doing my first live TV show appearance. So I'm puzzled by all of this and then, out of the blue at the end of the show, Master T says "We usually don't allow record execs here, but tonight is something real special" and it finally kicked in. I was going to be presented with a gold record, the first Canadian urban artist to go gold in just over 30 days. It couldn't have been a better moment because T had a hand in that, fighting behind the scenes for urban music, period. Not just for Canadian music to be played, but black music period. And now he's presenting one of his own with an award that he had presented to other more commercial mainstream artists before. But now he's finally being given an opportunity to give it to someone who represents him, his culture, and his country. Receiving the plaque on any other show would have been cool too, but the fact that it was on his show gave more credibility to me, and the bigger picture. That urban music is here to stay. So you could only imagine how being presented with an award from Master T, on his own show, held special significance, not just to me but the whole urban music scene in Canada.

Much Master T

One of my last moments of glory before I left Much came when I had an "In Da Round" segment featuring Redman & Saukrates. Collabos like this, with prominent Canadian and American artists recording side-by-side out of mutual respect, is what my gig was all about. So to hear that Kardinal had recorded a remix with The Wu Tang Clan or that Choclair and Kurupt were featured spitting lyrics together on "Memoirs of Blake Savage" was this career vj's wet dream. Finally, Canadian rappers were sloughing off their typically hyphenated Canadian identity crises, finding their own voices, and not trying to emulate American emcees.

Knowing that I've been able to do my part in exposing the Canadian talent pool is very satisfying. When I left Much I could see that our artists were finally getting the big picture. By establishing their own labels and signing publishing deals, it was clear that these hip hop-reneurs were now starting to gain a sense of empowerment. There's still a lot of work to be done to strengthen this industry even more, but I'm encouraged that our country finally has a commercial urban radio station to spin the music around the clock. The launch of Flow 93.5 FM in late 2000, as Canada's first all-urban radio station, should radically change the face of urban music in Canada. Milestone, Flow's parent company, was finally granted a licence after CEO Denham Jolly spent 12 years of his life kicking and cajoling the CRTC to take black music seriously. Before then, almost all urban music practitioners were completely reliant on campus radio and word-of-mouth to get fans out to their shows or to promote their records (most of which were only available at mom and pop record stores, as well as some of the larger outlets). This radio issue has always been a sore spot for me. I've always thought it was a sad state of affairs that a video network was the only thorough source for

Racial Diversity and Urban Music in Canada

urban music in this country. I remember putting together a demo tape for an urban radio show (it would later air on Z103.5) and I was told by a radio program director that the music on the demo was too "underground" for Western Canada. This demo consisted of tracks by Puff Daddy, Brandy, Mase, Montell Jordan, and others — all multi-platinum-selling artists. Huh? The launch of Flow in Toronto kick-started a movement that has seen more urban stations pop up across the country, most recently in Vancouver and Calgary.

In the final analysis, I see the industry growing exponentially as hip hop continues to dominate the charts. But there still needs to be a way to develop talent in this country (youngsters need to learn about the business of music as well as the art). We also need to start grooming emcees to perform as Canucks (we're not American; Scarborough's not Brooklyn). Maybe this will help change the fact that no domestic-signed artist has yet to make a big splash in America or Europe. The few that are doing well (Glenn Lewis and Deborah Cox) had to leave the country for that kind of exposure. As well, the Juno Awards still only have one rap category, which means that full-length albums are still going up against 12" singles, so there's still a struggle ahead. To make things tougher, major labels are re-evaluating the record industry overall with the major drop in music sales recently. As a seasoned veteran, my position in the marketplace is to pass on these nuggets of wisdom, in addition to developing artists on my own Serro label. Also, I'm new school enough to realize that DVD is no longer just a buzz word, so I'll be looking to filter some T-branded merchandise into the market. We fortysomethings don't have all the answers, but continued growth and hope is on the horizon.

Much Master T

ROOTS, ROCK, REGGAE

If there's one goal I can truly say that I set out to achieve at the Nation's Music Station, it's the promotion of Caribbean music, namely reggae and soca, on North American airwaves. Going all the way back to when I started as a VJ on *X-Tendamix*, this was my mandate.

Outside of groundbreaking '80s artists like Messenjah, Leroy Sibbles, Inspector Lenny, Chester Miller, Carla Marshall, and the Sattalites, along with newer school crooners like Snow, Lazah Current, Puppet, and Blessed, reggae music doesn't nearly get the love it deserves in Canada.

In Toronto, it used to be that you'd have to tune in to community radio shows like *Reggae Riddims*, hosted by Patrick Roots on CIUT 89.5 every Saturday night, or listen to *Reggaemania*, Ron Nelson's show on CKLN 88.1, if you wanted to hear West Indian riddims. Playing music for a niche market, *X-Tendamix* fit into the MuchMusic format as a specialty show — at least in the beginning. So reggae and soca fit that mould to a T. John Martin, the original MuchMusic director of music programming, was quite receptive to the music. John had traveled to Jamaica to cover the late great Bob Marley's funeral in St. Ann's. This was an experience that was dear to Martin and he shared that monumental event with many of us at the station. Also, the original *New Music* (a show that created the template for MuchMusic) had covered a variety of events in Jamaica, from the Sunsplash Festival to the Bob Marley Museum. My reggae play list at this time was comprised of artists like Third World, Bob Marley, Peter Tosh, Aswad, Shabba Ranks, and others.

Racial Diversity and Urban Music in Canada

During my first few years as a VJ at Much, the one constant question I'd receive from *X-Tendamix* fans was, "Why don't you play more reggae videos?" My answer remained the same over the years: very few reggae videos ready for prime-time airplay were submitted. Luckily, I was able to remedy this situation (to a certain extent) over the years, and *X-Tendamix* became the only show in North America on which reggae, soca, and dancehall-loving fans could see live performances from stars in this genre. Thankfully, Toronto had such a large Caribbean population, and that meant that local promoters like Jones and Jones, Kings of Kings, and Lance Ingelton Promotions (LIP) would consistently want to build up hype for the artists they were bringing into town to do stage shows. It was a win-win situation, really. I would get some of the biggest names in reggae music to appear on my show, and the artists and promoters would receive national media attention.

The mayapples who thought reggae music wouldn't crossover into the mainstream in Eminem-esque proportions are looking awfully silly now. Do you know what the biggest selling album in America was for 2000–2001? If you guessed Alecia Keys's *Songs in a Minor* or Creed's *Weathered*, you're way off. The answer is Shaggy's *Hotshot* album, which sold 800,000 in Canada and over 11 million copies worldwide, boys and girls. In the early days, I was unaware of the cultural impact my reggae segments would have on Caribbean communities across Canada. But just by having their A-list reggae heroes stop by, hang out, and oftentimes perform their hits live with Master T and Roxy by their side was monumental. Not only did this marginalized music touch all transplanted reggaephiles, it would also make a major dent on Canada's multicultural mix. Over the years I literally had every major reggae and soca artist perform or

Much Master T

hang out with me on the show. Everyone from Beenie Man, Bounty Killer, Shaggy, Super Blue, and Tony Rebel, to Freddie McGregor, Patra, Lady Saw, Sanchez, Elsworth James, Sean Paul, Mr. Vegas, Ronnie MacIntosh, Baby Cham, and Elephant Man.

I hold a close place in my VJ heart for Shaggy because he performed on the first reggae *Intimate and Interactive* show at MuchMusic. I was grateful to have been granted the opportunity to interview him because it provided a fitting end to my VJ career. I'd already interviewed him for shorter segments many times before, but for this *I&I* we were on a mission to prove to Canada how expansive and commercially palatable this music could be. Leading up to his performance, the whispers around the Much building implied that he was a one-hit wonder, which made me realize that reggae music was still not getting the respect it deserved. Naturally, the expectations were low, some of the staffers at Much didn't expect a large crowd. Approximately an hour before the show, the streets hadn't yet filled up to Madonna-esque proportions, so it was turning into an even tougher sell. By the time Shaggy and his Big Yard camp (comprised of Rayvon, Rik Rok, Tony and Brian Gold) launched into their rehearsal session it made believers out of everybody. These cats had a tight band and an even tighter group of back-up vocalists and dancers. In fact, just minutes before Shaggy's set I told him that a lot of folks still didn't believe in this music and that he had to recognize the significance of this show from my standpoint as a reggae booster. The rest, as they say, is history. His set was one of the most engaging *I&I*'s that I have ever been a part of.

In December 2000 I did an "In Da Round" segment with Shaggy and at the time "It Wasn't Me" was the number one song on the *Billboard* charts. It was an awkward performance because he did

Racial Diversity and Urban Music in Canada

three songs to DAT and, as a VJ, I was embarrassed. This guy deserved a much longer segment and he should have been given the opportunity to perform with a full band. In any case, I don't think the performance was the first thing on his mind at the time. Y'know those whispers between VJs and stars that you don't get to hear over the air and that make it into tell-all biographies? Well, let's just say that Shaggy had some powerful powers of prognostication because he was informing me, even back then, that *Hotshot* was gonna ship five million plus.

It's too bad that it took the Much execs until *Snow Job 2000* to catch on to what a great performer Shaggy is. He was one of the highlights of the whole event. The musicianship was there, he was riding the snowmobiles, flirting with women in the crowd, and he was hamming it up for a good chunk of the set (you might even remember him crooning how "It Wasn't Me, it was Master T"). He had the audience fully engaged and they were eating out of the palm of his hand. It's a shame that it took so long for Shaggy to leave an indelible imprint on everybody's mind.

Before my Shaggy coup in 2001, there was Shabba. For the record, the first reputable reggae artist ever to appear on *X-Tendamix* was Shabba Ranks. The story of how I became acquainted with Shabba is one for the ages. It was 1992 and I was at a club in Toronto. This guy walks over to me and greets me in patois. The guy had a whole bunch of street aliases that he went by (that's why I don't remember his name) and he claimed to be Shabba's friend. At this time Shabba was, hands down, the most popular reggae artist on the planet. Anyway, this guy asked me to shout him out on air because Shabba would be coming to Toronto in a month. I didn't know what to make of this guy's claims, but I shouted him out on the air anyway.

Much Master T

A few weeks later, Shabba's Canadian Sony affiliate brought him to Toronto for a promo stop. They brought him to the Real Jerk restaurant that used to be at the corner of Richmond and Duncan. I went by with my Much crew to introduce myself, and there he was munching on some Jamaican eats with the nameless guy I had bumped into at the club weeks earlier. I introduced myself and Shabba said something kind about me representing reggae music in Canada. Coming from the biggest reggae artist in the world, these words were humbling. Hours later, we headed back to Much and I set up our interview segment to be taped in the backroom area and aired later on. The one thing about Shabba that stood out to me was not necessarily his gold front tooth, but that he had this large entourage — and an even larger bodyguard who stood at least six feet, six inches — that he travelled with.

Before the interview, Denise Donlon had asked me to quiz him about some allegedly homophobic lyrics he had uttered a few years earlier, but I didn't bother to go down that route. He had already paid a heavy price for his ideas on sexuality, with primo talk shows from Letterman to Leno cancelling his gigs. Besides, he had explained his position on national TV many times over, so delving deeper into this scandal was dated.

Instead, I asked the king of dancehall to do a freestyle segment with me where he would toast and I would sing lovers rock–style in accompaniment. I would be the Tanto Metro to his Devonte. Amazingly, he did it and the segment was a success once it aired a week later. After this segment, my show received immediate street cred as a reliable source for reggae music.

As far as taking the largely underground dancehall sounds of Shabba Ranks into the mainstream, I programmed a couple of

Racial Diversity and Urban Music in Canada

shows that will go down in the Much reggae annals. In 1997 I hosted and produced an "In Da Round" that featured three of the biggest-selling reggae artists, Beenie Man, Buju Banton, and Sanchez. This show was huge for me on too many fronts. For starters, I had taken a two month sabbatical from Much to help my wife deal with the high-risk birth of our son, Kalif. And on April 5, 1997, just two weeks before the birth of Kalif, my father had passed away. In retrospect, I have to give Denise Donlon and Moses Znaimer credit for handling this sabbatical scenario the way that they did. They didn't have to let me take all that time off, it could have been business as usual for them, but it wasn't. From a personal, emotional standpoint, this show — my first since coming back to the station — was a rollercoaster ride of epic proportions.

You can imagine how emotionally charged I was for this show. I was hyped to be coming back to the job I loved, and I was looking forward to a great show. At the same time, I was still in shock about my dad's death. Ever since my brother Basil had called me on my cell phone to tell me that my father had died of a massive heart attack in New York, I had been devastated. But I had to keep moving forward. I knew he was with me.

The audience for this show seemed secure in their minds that dancehall was here to stay and that it would be a force to be reckoned with. It was inspiring to see the crowd singing along to Buju's "Wanna Be Loved" word for word. Although I was paying attention to the proceedings as the host and producer of this segment, my mind was also drifting. Paula had said she'd bring Kalif by the show, so I kept looking out for her in the background; my eyes lit up when I finally spotted them. I brought Kalif, who was only a few weeks old at the time, down to the studio floor, held him up, and introduced him to the nation. I'd end up bringing him by my show for every single one of his birthdays.

In the end, I dedicated (or dreadicated) this show to my pops and, boy, was this gig emotional. Of my 12 years in the business, this show was definitely the most cathartic. I had to prove to myself that I could push aside adversity of the most extreme form and still enjoy this VJ thing, still make the show exciting. Some of the Much viewers didn't get the significance of this show for me because they didn't understand what was going on in the backdrop to my life. I was getting voicemail from fans who just wanted to watch rap and R&B videos like they would on any other show. But this wasn't any other show. Reggae was finally making its mark; I had lost my father, who was my idol; and, with the birth of my son, I had become a father myself.

The following year, I achieved another major reggae coup for an "In Da Round." With the help of Kings of Kings, one of the largest reggae promotions companies in Canada, I was able to land Beenie Man, Bounty Killer, Little Kirk, Silver Cat, Richie Stephens, and local star Kiddie Ranks. This show was significant for a plethora of reasons. First, when local urban clothiers Jaydees pulled out some ad spots for the show, it meant the reggae community was starting to lend some fiscal support to our efforts. The word on the street is that this spot gave them upwards of $20,000 in business. Second, we had landed two of the biggest dancehall artists in the world, Beenie Man and Bounty Killer, who had already been carrying on one of their numerous verbal feuds for quite some time. In fact, during the interview segment, Bounty was taking verbal stabs at Beenie saying things like "I'm hardcore, I don't do pop." Beenie was the more diplomatic of the two, seeing as he was more interested in crossing

Racial Diversity and Urban Music in Canada

over into the mainstream (he had a hit with Mya and has recently worked with Janet Jackson).

Despite the excitement of having these two artists, it was difficult, as a producer, to juggle five different guests in an hour-long show. Thankfully, the DJs I had booked for the gig, the ever-popular King Turbo and Lindo P, played a wicked set and provided great segues for each individual performance. It didn't hurt that Beenie was so captivating, while Bounty's calling card tune "Look into My Eyes" had the crowd flashing their mock lighters in the air (this is a ritual a reggae audience does to support a song, the equivalent of North American audiences clapping for a good tune). We had to edit out some of the cuss words from Bounty before we aired repeat performances, but it didn't matter because the show had been a ratings success.

Other standout moments in my career programming reggae include interviewing Burning Spear, my fave reggae crooner, and having this young, white, prodigious dancehall singer named Snow come by my show and tear it up. I grew up listening to Spear's music as a kid. My father used to spin his records and I loved his mystical vibes. Whenever he'd come to town to do a gig, I'd go to his shows; I'd seen him perform at the Copa, Concert Hall, and the Warehouse. After his Warehouse concert we learned of a potential interview op, so I drove up to his Scarborough hotel to land this interview. I was nervous as hell because it was Spear, one of my musical idols. To make things worse, the contact phone number we had been given to reach him was not working. So I went upstairs to his hotel room to find him. You won't believe this, but he was locked out of his room and I had to interview him in the hallway. It didn't matter. Hallway, highway, or hotel room, I would have interviewed

Much Master T

him anywhere. At the end of the day, the interview turned out well.

And then there was Snow. Y'know, before anyone had even heard of him, this white guy from the projects in Scarborough was quite the talent. Rupert from Messenjah, a Canadian reggae band, kept telling me about this guy from his camp who was gifted at crooning reggae tunes. So one day I booked Messenjah to come on my show and there was Snow accompanying them. His a capella delivery was a real show stopper. You would never have known that this guy was, well, white. As a performer, his specialty seemed to be taking different reggae influences and blending them into his own delivery (some reggae critics say he sounds like Michael Rose of Black Uhuru, others say he does a good job of mimicking Pinchers and Tenor Saw). All I can say is that when Eminem is one of the world's most famous rappers, when Snow is Canada's best-selling reggae artist, when Venus Williams is the world's best women's tennis player, and when Tiger Woods is the best golfer in the world, you know the times are a-changin'.

One of the best experiences I had as a VJ was covering Sumfest in Jamaica, four years in a row. All the arrangements were always made weeks before we left Toronto. In the first two years, we hired my friend Michael to be our driver in Mobay. Then he turned us over to his friend Robert, who drove us during the next two festivals. After we arrived in Montego Bay, we were taken to the place where accreditation was completed (usually a hotel working with Sumfest). This process was always very thorough because of the magnitude of the festival and the degree of security surrounding the event, especially the stars. We were issued wristbands, which helped the security determine our access privileges to different areas backstage and in the pit. The Much crew was always assigned all-access

Racial Diversity and Urban Music in Canada

wristbands, so our passes allowed us to go directly backstage. To get there, we had to pass four checkpoints with police and security.

I did a bit of preparation for the interview process in advance of each night, but mostly it was a spontaneous go-with-the-flow and capture-the-moment vibe. That's what makes the whole thing so incredibly exciting. But this gig is exhausting from a vj standpoint as well. You arrive and set up at 11 p.m. to get the artists before they go onstage or after they finish their set, and your work ends between 5 and 7 a.m. the following morning. So you're working for about eight hours straight, outdoors and very late at night, looking to snag interviews, get clips, and so on. The fact that the fest organizers book the media in five- and six-star hotels helps ease some of this stress. The crew and I were put up in exclusive resorts such as Round Hill (where *How Stella Got Her Groove Back* had been filmed), Half Moon (where Queen Elizabeth and the royal family had stayed), and Tryall (known for its exclusive villas and a world-class golf course). On occasion we also got to rub elbows with the artists because some of the performers for Sumfest were put up in these hotels as well.

Each year, Sumfest delivers a full range of international artists; it's a real mix. In 2001, Beenie Man, Snoop, Ja Rule, and Blu Cantrell performed on Dancehall Night; on Legends Night, Cocoa Tea, Freddie Mcgregror, and Tony Rebel highlighted the event; and on International Night, Shaggy and Damien Marley flexed their muscles. But it's near impossible for any one artist to steal the spotlight at this type of festival. The Sumfest crowd makes the audience at the Apollo in Harlem or some audiences in Toronto seem easy to please. They're tough and they take no guff. One year when KC-i and Jo Jo performed, the folks behind me were just tearing up those poor boys. They weren't used to that type of response and had to buckle

Much Master T

down and turn it out. The American acts have to prove themselves even more on Jamaican soil. Sumfest also attracts diverse crowds with international flavour. People fly into Mobay from all over the globe to enjoy the festivities, while the locals come out in droves to support their favourite stars. It's a Red Stripe at 11:30 p.m., a rum punch at 2:45 a.m., and a Blue Mountain coffee at 5 a.m. You've got to pace yourself.

In retrospect, I wish I had been able to program some more obscure Caribbean musical forms (like zouk, compas, and calypso) but I was thankful for the opportunity to book at least a few calypso artists during my tenure. It was frustrating for me because I wanted to support the soca scene more, but the videos were largely low-budget affairs shot on video. Machel Montano was one of the few soca artists who had crossed over. He had videos rotating on Much, so I could take a chance with him. He came on my show once, did a live set, and tore it up. I relied heavily on my Much crew to be a good gauge of whether booking these diverse acts would work, and they thoroughly enjoyed his set. The other group I really loved and had to make a part of my show was Alison Hinds and Square One. Hinds had come on *Da Mix* and performed "Faluma," winding up her waist to the delight of my viewers. When I needed to book a soca act for my 10th anniversary show, I called them up again. They were the consummate professionals. They understood the value of having a good video and marketing their music, so whenever they'd come to Toronto they would use Canadian production facilities and shoot a slew of videos. I sincerely hope that other soca and calypso groups will follow their lead so that soca acts will, one day, receive regular booking on MuchMusic's new weekly urban music show, *Tha Down Lo.*

Celebrity Interviews

CELEB HOPPING

It's quite the task to choose highlights from amongst the 200 or more celeb interviews I did in my 12 years on the air. Actually, I've been so frequently bombarded by curious fans wanting to know the 411 on their pop heroes that Paula suggested it might be a good idea to put together a section that highlights some of my experiences. "Hey Master T," people continually ask me, "what's so and so really like?" Now, I've always felt that I was, in some way, a special link between some of the greatest artists in the world and the rest of the nation, but after being tapped for celeb gossip too many times (and in the middle of too many private moments — a bathroom stall is not the best place to ask about a celeb, hello!), I knew I had to go public with some of my insider insights into celebrity mystique.

Much Master T

Coming up with a list of noteworthy interview subjects to wax about hasn't been easy. Just because an artist has reached pop icon status doesn't mean there'd be a good story to share. So only the very best interviews and the most interesting characters — regardless of an individual's celebrity status — have been chosen for this particular ride. What was it like to practise breathing techniques with Madonna? How did I enjoy dance lessons from Britney Spears? What did I think when Scary Spice pulled my head into her cleavage? How did I like flirting with Janet Jackson? And what did I think about being firmly reprimanded by the love God, Barry White? Read on . . .

John Hiatt
(September 12, 1993)

My first and only interview with blues rocker John Hiatt has to go down in the T archives as, hands down, my worst interview ever. One would think that, after interviewing more than 200 entertainers over a 12-year period, it might be a challenge to find one standout stinker. But it wasn't. By the time I left Much, I had developed a reputation as a VJ who did a fabulous job of researching his interview subjects and boy, do I have John to thank for that. Why? Because he showed me the kind of mediocre results you can get when you do an interview-by-the-numbers routine and completely wing it.

John had come by the Much studios to promote his new release, *Perfectly Good Guitar*. He had just turned 41, and *Perfectly Good Guitar* marked a new direction in his career. It appeared as though

Celebrity Interviews

he was looking for a hipper, more rock-oriented sound (the new album had been produced by Matt Wallace of Faith No More and featured Michael Ward from School of Fish). I didn't have time to research his career properly and I wasn't very well acquainted with his music. As a seasoned veteran in the industry, he caught wind of my ignorance and quickly made mincemeat out of me . . . in a diplomatic way. For starters, I opened up the interview by fumbling the name of his release, which isn't exactly the best way to get a ringing endorsement from any celeb, much less John. His band came by to do a live in-studio performance, and after he introduced his band, I started firing off a string of dull, ad-libbed questions that could only come from a VJ who knows very little about his subject's music.

I started out by completely ad-libbing my first few questions. Stuff like, "Uhhh, what's it like turning 41? Are you going through a transitional period?" Then I bluffed my way through the next couple of questions, prefacing what I had to say with "from what I've read about you." It was obvious that I hadn't read much about him. By the time we got around to talking about how his new record was inspired by his 15-year-old stepson Rob, he had become the interviewer, and I had become the interviewee. I used a word out of the '90s hip hop slanguage to describe the type of sound he was going for (the word was "fresh"), and he had a field day with that one. He started to string off a few of these hip hoppy terms like "fresh" and "phat," as if to mock me. And then he said something along the lines of "I think it's a good time for music, don't you," quizzing me, instead of the reverse. Straight up, whenever your subject starts dissecting you, the interview is a write-off.

But that didn't stop me. I fudged some more, making non-specific comments like "you've obviously influenced a lot of bands coming up" and "this record for you is like a coming out." In response to that "coming out" question, John screamed, "I'm out of the closet, baaaaby!" At that moment I wanted to hightail it out of there. Seemingly bored and uninterested, John then asked his label rep for a towel to wipe his sweaty brow — literally interrupting me mid-way through a question. He appeared flippant, distant, and detached. In my mind I knew he was looking at me and thinking that I was unqualified to be conducting interviews with an artist of his stature. And boy was he right. At this point, he probably just wanted to play his tunes and get the hell out of the Much environs.

I made a promise to myself after this not-so-fine day. Never again would I go into an interview (even with the smallest indie band that nobody had ever heard of) without researching my butt off. My learning curve as a VJ shot up from this point on, as I realized that I needed to broaden my musical horizons just a tad if I was going to interview non-urban musical artists. From that point on, I'd never go into any interview without at least reading my Cliffs Notes. John's band closed the set with a song called "Blue Telescope," and I ended the interview by crawling back into my hole.

Bobby Brown
(X-Tendamix in New Jersey November 14, 1992 and November 22, 1997)

If you've watched any music awards shows over the last few years, you've probably seen Whitney Houston claiming that Bobby Brown "is and always has been the King of R&B." Well, I'm not so sure about

Celebrity Interviews

these biased pronouncements. There was a time when Brown, the youngest member of New Edition, was a respected entity in the R&B game. His second album, *Don't Be Cruel* (1988), had made him one of the biggest R&B stars of the '80s. But what had he done lately?

I was scheduled to interview Brown in New Jersey at his home studio in 1996. When I first heard about the interview location, I was curious and excited that I would get an opportunity to see, not only how Bobby was livin', but especially how Whitney was kickin' it. Bobby was reported to have this huge studio (which, I later found out, he referred to as his "baby"), where he was in the process of producing his upcoming album, *Forever*. For some odd reason, when I was told that the studio was in the house, I took it literally. I had pictured myself interviewing Brown in a room inside their palatial estate. But when we exited the car in the mid-afternoon, they received us and showed us to the place in question. We were actually escorted to another house, a three-bedroom bungalow with a large modern kitchen that had all the amenities. This house was home to the studio.

Brown came out and we got re-acquainted. He was chilled, real cool, like one of the guys. He gave me a tour of the studio. What's really sad, on my part, is that the whole time he was giving me this tour I couldn't stop thinking that Houston must have paid for all this. There had been talk about him going bankrupt and having major financial problems. As part of the tour he showed me the trophy room where awards as well as gold and platinum records were on display. And like all celebrity digs, they had a happening game room in the crib as well. He explained that they like to entertain and party there quite often and he liked to cook up a storm in the kitchen. He also mentioned that the neighbours would complain about the noise

Much Master T

on occasion. After the tour, we settled in and got all the technical stuff out of the way; we were ready to hang. I began the interview by jumping in and asking him about the New Edition reunion and why he thought it didn't work out. He was quite blunt. He told me that some of the group members were jealous when he'd do his solos during the live shows. He claimed that when they were touring and the audience would be more in his favour, that the other group members weren't feeling the love as a result. He also implied that he was getting jerked around financially too, as far as the money splits between group members were concerned.

I was slightly on guard during the interview 'cause I was hoping that Whitney might pop in for something, anything. Maybe she was feeling a little nippish and she'd come to the studio to partake in the lovely spread they had laid out for us; maybe a kiss from Bobby was on her agenda. It would be a coup just to talk to Whitney Houston. The Browns/Houstons had an intercom set up between the houses and Bobby was having a conversation with his daughter, Bobbi Christina, as we were doing the interview. She'd be saying, "Hi Daddy" over and over. At one point I felt like jumping in and saying, "Bobbi Christina, it's Master T here. Can you go get mommy, pleeeeze." But that never happened.

When the interview was over, I was looking out the window in the kitchen. As I was taking in the magnitude of the mansion just above the trees, my mind jumped to the first time Bobby Brown had come on *X-Tendamix* in 1994. He was the biggest recording artist I had ever had on the show at that time. It was a walk-on segment and we hung out with Roxy. He was a real gentle soul back then, still is. But he's slightly different now. He wears some of life's experiences on his face. Back then we talked about different ways to cook chicken — he's quite a

Celebrity Interviews

gourmand. Personally, I think the whole bad boy image is something of a media creation. He made himself into a victim by getting caught in the limelight. When his star started to dim it would have been a nice to see him re-direct some of his energies toward something else, like becoming an artist's manager or a producer. But it's obvious that he just has one of those personalities where he craves the spotlight and the superstar perspective. To see him tagging along with Whitney at the awards shows, as she blurts out that he's the King of R&B, is embarrassing, to say the least. It would be wonderful to see him capture the spotlight on another level; he's still got loads of talent. But who am I to say. In any case, it's his prerogative.

Barry White
(November 30, 1994)

Barry White is the god of love. Pam Grier is the goddess. What kind of lovelorn fool wouldn't want to interview and hang out with his Majesty, Sir Barry himself? When Michelle Geister, the producer of *Soul in the City* at the time, asked me to interview Mr. White in 1994, I wasted no time in accepting. The occasion was Mr. White's big comeback album, *Barry White: The Icon Is Love*. During his time away from the biz, he had completely revamped his image by losing a lot of weight and visually streamlining his look. This was also a time of change for me. Paula and I were in the process of closing down T's Cribb, due to the numerous break-ins we had experienced at the store. And Michael Williams had left the building.

Prepping for the interview was exciting but difficult. Barry was

Much Master T

definitely a legend, and (from what I had seen of him in other interviews) he seemed to have a bit of a firm disposition about him. I knew I would have to step to the plate and come correct to gain his respect. Barry was only doing one interview while in Canada, so I certainly didn't want to embarrass myself and ruin the Much exclusive. The segment was to be aired as part of a one-hour special on *Soul in the City*, but in my twisted mind, I had a plan to kill two birds with one stone. *X-Tendamix* was my baby at the time, and Taurus T was one of my crazy cousins who idolized Barry White. I knew I couldn't miss out on the opportunity to have my show blessed by the love god himself.

Leading up to the interview I felt more confident than usual 'cause Basil was shooting it. My best friend was also there, Dave Campbell, *Da Mix*'s in-house DJ, who had brought in some old, dusty vinyl records for White to autograph. The A&M label rep seemed perplexed as to why three grown black brothers were acting like silly putty in this man's presence. Maybe it's because Sir Barry himself is responsible for about one in every 20 babies born in the '70s. The interview was to take place at the Top O' the Senator (a legendary jazz bar in Toronto) and Barry showed up a little late. When he finally walked into the room, he was dressed in black with a matching trench coat. He stood about six feet, four inches and his voice was even bigger. White was immaculate, clean from head to toe, and groomed to perfection. He had that classic, old-school vibe the players used to strut. It was funny to hear my brother Basil telling White how many times his music had helped him get to third base with the ladies. Somehow it had come out in conversation that Basil and I were brothers. He gave us some good words of wisdom regarding what it's like being black in the industry, imploring us to not let the industry tear us apart, and to stay

Celebrity Interviews

tight. He said that he was surprised to see two young black siblings working so closely together in the media. He definitely seemed pleased with the Young brother combo.

When we sat down and began the interview, I knew I was there to represent *Soul in the City*, but deep down inside, I couldn't stop thinking about how I was going to convince Mr. White to do another small segment with me acting as my Barry White–influenced alter ego, Taurus T. So after the interview I told him that I do this Taurus character. I tried to explain what the meaning behind this character is, how, conceptually, Taurus T is his love disciple, blah, blah. He just looked at me as if to say, "You've got to be kidding, don't you know who you're talking to?" Next move. I asked him to do this improv comedy sketch with me. He glanced at me and said something along the lines of, "I don't mind doing things like this, if I'm properly informed. Y'know, we as artists get a bad rap when we say no to things like this. You should have faxed me over this request so I could have read it over to make sure I knew exactly what I'd be doing and what I'd be saying, come on now!"

By this point, I was feeling like an absolute nimrod for making this request and taking all these potshots from Mr. White. But I was very unapologetic, after all, he is the love god, so I wanted him to give this thing a chance. For some reason I had a feeling that if I hung in there long enough and suggestively let him mull it over, he just might break down and let it happen. Which he did. When he finally gave in I got right into costume, putting on Taurus's shirt, sunglasses, and the like. At this point, White was looking at me and saying to himself, "What's this brother doing?" But I was driven by that other God to pull off this sketch comedy piece for my *X-Tendamix* viewers. Let's face it, opportunities like this only come around once in a lifetime. So I looked over at Basil and asked him if he was ready to roll. Basil, standing there relaxed, sensed how frantic I was and quickly prepared for the shot. The minute White heard my Taurus T voice it was magic. We did it straight impromptu, no editing or strategically placed cameras anywhere; the whole segment was ad-libbed in one take. Magically, we had managed to pull it off. When White left he said, "You Young brothers have definitely arrived."

I never made a habit of repeating segments on any of my shows, but we must have aired this one at least two or three times. If I had to rate the interview on a 10-point scale, I'd say it was an eight, but the snippet was definitely a 10-plus; it truly was some of my best comedic work to date. I called Paula right after the shoot and told her of the coup. We hooked up later on that evening to watch the tape in an edit bay, our infamous after-hours ritual, and she lost it. She was bursting at the seams. She must have watched it three or four times that night. A few days after that monumental shoot, I walked into the Much Environment and there was this crowd standing around one of the edit bays, tearing themselves up with laughter. The news of my risky endeavour had spread, and I was never so proud to have stooped so low . . .

Luther Vandross
(February 15, 1995)

Both Paula and I are huge Luther fans. As you might remember, there was a debate over whether to play Luther or Sade as the special song at our wedding and Sade won out, but just barely. Hooking up with Luther was to be my first interview with a bona fide star, so I

Celebrity Interviews

was on a mission to do all Luther fans proud. Paula and I were going over questions for about four days prior to the interview to make sure I had all the angles covered. This was to be a closed-off, sit-down interview. The record company had sent over a rider to make sure the room was the right temperature (no air conditioning), and that all of his requests regarding towels, juices, etc. were met. I was worried that he might turn out to be a total prima donna, but my fears were put to rest early on.

For my Vandross exclusive, literally every person of colour in the Much building was circulating around the boardroom waiting for Luther. They wanted to use me as a conduit to get autographs from him. Unfortunately, it's every fan for himself. I rarely cross that line (and when I do it's to get something for myself), but I don't even do that until after the interview takes place. My brother Basil, another big Luther fan, was shooting this interview as well. Although I was feeling a touch more nervous than he was (for obvious reasons), we share the same relaxed intensity in these circumstances. It's all about focusing and capturing something extra special. It was early afternoon when we began the interview process in Moses' boardroom. When Luther arrived I was sitting down and getting myself organized. He was near the door getting acquainted with the environment. For whatever reason, I could tell that he was checking me out, looking me up and down; taking in my jewellery, my locks, my attire. It was a slightly uncomfortable moment, but I understood that this was Luther: his star status was on a whole other level.

At this point in his career I could visibly see that he was in his medium range, as far as his weight was concerned. He was wearing a pair of fitted jeans, which was unusual for Luther (we later discussed this new casual image and he seemed surprised that I had noticed

Much Master T

that he was sporting a whole new look). When he sat down, Basil fitted him with a microphone and we prepared to commence the proceedings. But just when I looked up from my question sheet to make eye contact with him and get the go-ahead, my focus drifted to a gaping area of Luther's shirt. It had come unbuttoned and Luther's stomach was visible through the shirt. My eyes were slightly fixated on it and I was pondering exactly how to handle the situation. To make things worse, it was quite obvious that he was wearing a stomach tensor. Oddly enough, I was somewhat familiar with this type of undergarment because my Uncle Lloyd used to wear one all the time in New York. This was distracting at times and a part of me wanted to let him know about it, but obviously I didn't know him well enough to broach the subject. Where does one begin? "Say, hey Luther, uhhh" . . .

As the interview got into full swing I finally stopped focusing on that little situation with his shirt. Luckily, I was so wrapped up in the conversation that I was able to get my mind off it. I was just praying that some of the footage would be salvageable. Hopefully Basil had gotten a lot of close ups. After the interview was over and we were wrapping things up, Luther and his entourage departed. I motioned to Basil and we spoke briefly. Apparently he had also caught Luther's little exposure and told me that he had consciously made an effort to shoot the majority of the interview from an angle that didn't show the problem. In the end, I felt that Vandross had enjoyed the segment and I think he appreciated how well prepared I had been for the gig. More than anything I was very pleased with the connection I had established with Luther on camera. I think this enabled us to bring the viewers a solid, heartfelt interview.

As I mentioned earlier, this interview had a rider. The rider clearly outlines what you can and cannot do around Luther. We were not allowed to talk about Luther's sexuality (I'm sure you've heard the

Celebrity Interviews

rumours). Bearing this in mind, the interview was quite revealing for the passionate Vandross fans out there. We learned that Vandross has perfectionist tendencies and that he would never settle for mediocrity. On camera he told me of a situation that had arisen when he was touring with En Vogue in 1995. Apparently the R&B quartet had originally signed some contracts to tour with him and the rules and guidelines for the tour had been clearly laid out (there were certain stipulations that they had to comply with). But as soon as the tour commenced things changed. There was a big hoopla at the show in Toronto because the two camps weren't getting along. He explained to me that En Vogue had breached their contract in a number of ways. You could see the disgust on his face when he told me about En Vogue lip-synching to DAT, which he felt compromised the whole credibility of his show. When you go to a Luther show, it's all about turning it out live.

A number of different people have pointed out to me that I have a special skill for manoeuvering around tough questions that might make an artist uncomfortable. One of my secrets is to tap into my God-given sense of humour. We didn't manage to get any sexual dirt on Luther in the interview (and I knew not to go there), but we did come up with a Top 10 segment on things you need to know about Luther. For example, Luther's top vacation spots, fave meal (he loves McDonald's and Mrs. Field's cookies). Arguably the tastiest tidbit we uncovered came when I asked him how much he had ever spent on a single shopping spree. His reply was $30–40,000. Despite the early feeling out process, Vandross became more comfortable as the interview progressed. Eventually he started to share more of his life with the viewers. He almost seemed disappointed when it came to an end.

Paula and I edited this interview ourselves, so it was a labour of love. Paula would come in to help edit because I didn't have an associate producer at the time (which definitely helped me manage my workload). A lot of the editing was done after work — after Paula had carried out her design duties and I had put in eight hours producing the show. It wasn't uncommon to find us there editing material into the wee hours of the morning, sometimes until 5 or 6 a.m. Many times we'd be leaving as the *Breakfast TV* staff would be coming in. For the *Luther Vandross Special* (this format was the precursor to the *Chillin' With* shows that I later established), we combined rehearsal footage from his tour with the interview that had been shot by Basil, strategically overlaying background music (all of it was from Luther's new album, of course). I programmed the show to include a variety of Luther's videos and packaged a stellar special that was aired to rave reviews. I wanted the segment to speak volumes about both Luther and my talents as a producer, which I think it did. Up to six months after the special aired, fans were still coming up to me and recalling the interview's finest moments; they'd tell me how it inspired and touched them. I ran into Luther at the Soul Train Awards the following year. He had won the Sammy Davis Jr. Lifetime Achievement Award. When he remembered me, I felt honoured that I had made an impression on him. Paydirt.

Mariah Carey
(May 2, 1996)

Without a doubt, my interview with Mariah Carey is filed right atop my list of least interesting interviews. And it happened on my 34th birthday. Who could have asked for a better present, right? It wasn't so

Celebrity Interviews

much that I sucked or that she was dull and uninteresting, it was the controlled karma that sealed this interview's fate. Compared to my Madonna moment years later, there seemed to be a touch more nervous tension in the Much building as we went into this interview. Carey and her label reps at Sony had a lot of demands. When she arrived at the Much building, she came with an excessive number of (now how shall I put this?) puppet master-like assistants. She was still married to Tommy Mottola at the time, so maybe that explains the situation (it was alleged that he dictated everything about how she was to be perceived by the public). Before I could even start to research this interview and figure out an angle, Carey's PR folks told me that she wanted to know, in advance, everything I was going to ask her in the session. This meant that whoever participated in this show (producers, etc.) had to come up with about five questions each for a total of 20 or so. She would then tell us the questions she found appropriate. As a VJ who used to enjoy the edge and the spontaneity that comes from thinking on my feet, this really took away from the whole experience.

Sadly, we had also been given instructions not to open the windows. For some reason, they just didn't want the fans to see her up close. If you can imagine this, there were tons of fans gathered outside the Much environs and, when she entered, she said something like, "Oh, look at all those people outside." For whatever reason, she wasn't interested in having fans and autograph seekers jeer, scream, and whistle at her — all of which is fair game when you're a multi-platinum-selling artist. I wanted to say to her, "Well, these are the fans that pushed you to platinum. Hey, if we open up the window, then maybe they'll be able to feel closer to you, more connected." But I just had a sense that she wasn't gonna go for it.

But that wasn't the only issue with Carey. If there's one thing I

Much Master T

don't like to do with interviewees, it's sit down with them, hang out, and make small talk before the interview. Personally, I think it detracts from the freshness of the conversation. Carey's manager requested that I rap with her for 20 minutes pre-interview so that she could feel a little more comfortable with everything. I knew I didn't want to sit down with her then. What were we gonna talk about? How controlling her husband is alleged to be? The weather? How about those Leafs? It's not like I thought she was an airhead or anything, I just didn't feel like I'd have enough to say to her outside of the interview material, and this, more than anything, can be very awkward for me. It was nothing personal.

When I finally got up close and personal with Carey she was looking good to me, but some of female staff at Much quietly expressed that she was looking a bit on the heavy side. Some even went as far as to say she looked pregnant at the time. Interestingly, a tabloid had just reported that she was mysteriously peeing a lot (that's a common symptom in the first trimester of pregnancy). Even in the middle of our interview she went to the ladies room with three of her assistants. You read that right. What did they do? Hold the door shut for her? Maybe pass her some toilet paper under the stall door? I have to admit that my tabloid-o-meter did go off when she made that move. Anyway, once Carey got settled in, away we went. I was working it, but for some reason (maybe I needed a breath mint, who knows) there was a certain disconnect between us. I asked her, "Who does the cooking at your house, Tommy Mottola or you?" and her responses were short and abrupt. "No, it's Tommy," she said. She only had "yes" and "no" responses to everything. She was really guarded with her answers in this session, which came as no surprise. When I was sitting down with her earlier, I remember

Celebrity Interviews

hearing her make-up people telling us that we couldn't shoot her left side. Apparently it was her "bad side." I'll admit that I don't know a whole lot about make-up, but Carey was also the first person I had ever seen applying powder to her cleavage. I didn't realize that viewers could tell if one's cleavage appeared shiny on camera. I also remember asking her to sing a little something to commemorate my birthday, and she flat out refused. Instead, she opted to perform this whistle routine with her hands. Well, that was kinda cute.

At the end of the day, I think I was put off because I'm not a big fan of pretense. Having five people fussing around her and catering to her every whim left little time for her to give a bit of attention and a little energy to the desperate fans outside (all of whom had been waiting out there for hours, by the way). Don't get me wrong, the Carey today is not the Carey of 1996. You can see that she's grown a bit, finally asserting her independence from Mottola and I do believe that the recent *Glitter*-fuelled breakdown will definitely change her in some way. I see the disconnect we experienced in the interview as a possible effect of the Mottola control thang. What's the sense of doing an interview if you don't want your fans to know anything more about you? Now that her life and her image have been made over, I'd love to interview Mariah Carey again.

(MTV Awards, September 4, 1996)

What was it like to interview Tupac, the "ghetto Elvis," arguably the biggest icon in rap music (especially days before his untimely

Much Master T

death)? Well, let's just say that there's a Tupac in all of us. Tupac represented the schizophrenic duality or W.E.B Duboisian "double consciousness" that plagued the "Me Generation" of black males. In one breath he'd record beautiful odes to his mother ("Dear Mama") and black women ("Keep Your Head Up"), and in the next breath he'd voice some of the most irresponsible, vitriolic diatribes against everybody on the planet. One can also place Tupac in the tradition of Dead Prez and Mos Def with his intelligent and informed diatribes against The Man. He wanted to ensure that rap music wouldn't be completely dominated by licentiousness and mass consumerism. On the other hand, you can place him in the tradition of NWA, the group accused of setting back struggles for black equality by two or three hundred years. Scathing indictments of cops (otherwise termed "pigs," "Babylon," "beasts," and "boys in blue") — as well as raps about America's genocidal origins — either took you several thousand notches down on the evolutionary scale or provided hope, depending on your perspective. What could be more captivating to youth than the sight of a tattooed black male riffing anti-capitalist, anti-hot dog, anti-everything lyrics and bemoaning the depravity of "Babylon"?

I was fortunate to conduct the last TV interview with the late Tupac Shakur. This short segment took place just three days before Tupac was shot to death on September 7, 1996. As the story goes, he was riding in a car driven by the controversial Marion "Suge" Knight (who headed Shakur's former label, Death Row Records). The pair had just left the Mike Tyson–Bruce Seldon boxing match, and they were apparently en route to a nightclub. Knight survived the shooting; Shakur died six days later in a Las Vegas hospital. Like many in the hip hop community, Tupac's death saddened me. In the short time that I was

Celebrity Interviews

able to talk with him, I could see that he embodied the fire, passion, and angst that black men carry around with them daily as a result of the institutionalized injustices of the system. I wish *Rap City* fans could have seen the other Tupac, the good Tupac — not just Tupac the thug. Tupac was a gifted renaissance man-child. He consumed literature by Nietzsche and Nikki Giovanni; he was well versed in classical music despite the fact that he never graduated from high school and he spent a lot of time hanging out on the street corner. During this interview, Tupac had veins bulging out of his forehead and neck. He spoke with such passion that I would have dropped the mic and ended my career right there if I had known that this was gonna be the last time the world would hear his words in real-time. Here's a transcript of the interview:

Master T: Can we talk about Death Row East? What exactly is happening? What can we expect from it?
Tupac Shakur: Do you believe in God?
Master T: Certainly!
Tupac Shakur: Then believe in Death Row East, believe in that for real. If you believe in God, believe in Death Row East. We plan to take the same strategy we used with Death Row West, which is mind over matter; taking all our weaknesses and making them into our strengths and numbers. We already run these streets out here. So now we gonna just help some of these brothers get their money on. 'Cuz we know they got talent. We got the ways to make them use their talents to the maximum effect. And that's what it's about. Everybody raps. We don't rap. We rap to make money. We do business. Ain't no other record company out there that sold as many records as we did. We outsold Bad Boy,

Much Master T

LaFace, every black record label out there. We outsold them in one year! And I'm a convict and my homeboy just got off a murder case. So that tells you, it's pure talent. No hype, we don't got no all-around American smiles. They don't even wanna buy our record, but they gotta buy our record 'cuz we represent the streets. So Death Row East is gonna be a personification of what we did on the West side. We gonna do it to the East side. We gonna prove, once and for all, that all these people talking bout a East coast–West coast war, they like Judah was to Jesus. They only here to cause confusion. We here to bring money and to bring change. They here to bring confusion. All these weak rappers — Nas and all these suckaz — they battling off East and West like it's a game. This ain't no game. If this was chess, we'd be yelling checkmate three muthafuckin' years ago. 'Cuz we been beat these muthafuckaz. It's not a game. We out here trying to help out people make money; we trying to get out of this three-strikes circle they got us in and start getting our paper on. So that's what we doing. By strength and numbers we coming to the East coast to prove there is no fear, there is no problem, there ain't nothing but opportunity. And opportunity is overthrow the government y'all got right now, which is Bad Boy and Nas, and all that bullshit and we will bring a new government right here that will fear every person in New York.

Master T: Alright, man. Thank you. Take care.

When I had the opportunity to interview Nas, I asked him how he felt about Pac being murdered. Despite the warring words between the two, there was no real beef. Nas saw Pac in much the same way as he saw himself, as a real soldier. It's terrible when the hip hop press

Celebrity Interviews

blow these West coast–East coast stories up into something they're not. The beef between these camps existed more in the minds of magazine editors than it did between the artists themselves, or so it seemed.

Spice Girls
(January 23, 1997, and July 12, 1998)

The third time I interviewed the matrons of Girl Power, I was fully indoctrinated into their "Scary" Spice World. It was for a July 12, 1998, *Intimate & Interactive*. The Spice Girls were just on the brink of fame in America, but they were already massive in Europe. They had ushered in this Girl Power phenom (which was really nothing new considering that girl groups like Salt-N-Pepa and TLC had been singing about female empowerment through sex for years before). Still, the Spice Girls were on a whole other level.

Virgin Records Canada insisted that I be the VJ to interview the Spice Girls. I guess they figured that, because of my universal appeal and my so-called status as a seasoned vet, I might be able to corral these gals into spontaneously spilling their guts about anything and everything having to do with their newfound fame. Apparently, this exclusive interview was a coup; the only reason they consented to do it is that they had to make up for an autograph appearance they had cancelled the previous February. We were lucky to be able to catch these gals in the middle of the North American leg of their Spice World Tour.

The Girls had sold 31 million records worldwide (two million in

Much Master T

Canada alone), so it was no surprise that, by the best estimation of the Much brass, there were about 500 folks (mostly young girls) lining the street to catch a glimpse of their heroes. We limited the studio audience to about 100 people because things would have gotten out of control if we had filled the building to capacity. To the delight of rabid fans, Denise made the decision, shortly before show time, to open the studio window so that the Spice Girls could go out to the street and converse with their fans. Every sane female hanging on Queen Street that evening, regardless of size, shape, age, or race, wanted to be a Spice Girl. This *I&I* was one of the most enjoyable shows I ever did in my 12 years at Much. Imagine trying to control an audience of 6- and 7-year-old kids. Because of space restrictions, adults aren't really allowed on the set with the kids for these pop deals. The parents are escorted to another room in the building where they can watch the whole event on TV.

Even though they catered to a very young audience, the Spice Girls were going through some growing pains. They were maturing, right before our very eyes. Posh Spice was engaged to David Beckham; the English football star was in attendance outside of the camera sightlines. Mel B was engaged as well (though she's not anymore). Baby Spice and Mel C were allegedly suffering from depression and eating disorders. At the time, they were also being referred to as the Fab Four because they had performed without Geri "It's Raining Men" Halliwell at the Molson Amphitheatre the night before. So I guess all wasn't well in Spice World.

Although I did some prep work for this interview, I wanted to allow the proceedings to follow their own flow. I wanted the unique personalities of the Spice Girls to outshine any rigid, robotic interview protocol. The show began with Sporty, Scary, Posh, and Baby Spice

Celebrity Interviews

launching into their hit song, "Wannabe," which they sang live accompanied by a five-piece band. The next song they sang was "Say You'll Be There," and after that they did a rendition of "Spice up Your Life." In the middle of the track, the Spices went out into the audience and pulled a couple girls up on their feet to join the fray. The kids looked either too starstruck or too young to know exactly what was going on. After "Spice Up Your Life" they caught their breath and fielded questions from both myself and the audience.

This whole experience gave me a bit more respect for pop music fabrications. When a girl from the audience tried to ask the Spice Girls a question, she was so overcome with emotion that she burst into tears. She was quickly comforted by both Emma and Mel C. It really showed us a different side to the group. They were caring, compassionate young women who didn't seem to be too caught up in the trappings of it all. In a similar gesture, a fan who had camped out in front of Much for a day before the show was rewarded with the chance to kiss the hand of Mel B. In another classy (but strange) move by Mel B, she pulled a child out of the audience, held her on her lap, and started playing with the child's hair — all while the interview was going on.

Still, that was nothing compared to what she did on the Spice Girls' first visit to Canada. This was the legendary interview where I was verbally sparring with Mel B, trading licks back and forth. By far, she was the most animated of the bunch. I was telling her that I'd call security on her and, for some reason I was talking about my abs. At one point, they all started begging me to show my six-pack on the air, but I outright refused. Mel B apologized for being pushy and then, as she was making up with me, she got up to give me a hug, grabbing my head and burying it in her cleavage. Jaws dropped at

Much Master T

this point because people couldn't believe what they had just witnessed. I didn't do anything (honest). I was simply assaulted by two sizeable, brown, round, supple, Brit breasts (I still have flashbacks to this day; I may require professional help). The million-dollar question asked by the Girls' male fans: what did it feel like to have my head buried in a Spice Girls' breasts? Well, they were glistening and they smelled real good, all perfumy and stuff. Most importantly, they didn't feel all silicone-like; they may have been real. But that's a different story altogether.

Contrary to some reports, I didn't have a crush on Scary Spice. She just happened to be from my hometown of Leeds and she was black, all of which makes for great TV. A lot of North American artists tend to hold back, but the Brit sensibility lends itself to letting it all hang out, literally. To be quite frank, I was more focused on my wife Paula being seven months pregnant (and perhaps a little put off by the Scary attack) at the time. But Paula and I watched the interview together and she didn't think anything of it. With all her hormones out of whack, she told me that she hoped I had enjoyed myself 'cause that wouldn't have been happening right then with her.

The subtext to the fun and games that went on whenever the Spice Girls were around was the controversy surrounding Geri Halliwell. When I asked the uncomfortable "Where's Geri?" question, the room fell silent. Victoria fielded the question and said that there's no dirt to be had there. The band expressed disappointment over finding out about Geri's departure from the group just two hours before a scheduled performance in Oslo, Norway. It was a pure exercise in diplomacy on their part, as I got the sense that some of them wanted to rag on her. Mel C and Victoria seemed bent on clearing up any misconceptions about the group's identity.

Celebrity Interviews

It was funny because they were forthcoming with other bits of personal information. I got them to divulge info on their dream dates and their female idols. I asked them some thought-provoking questions like, "What other kind of spice would you want to be?" and they fired back with some witty responses. The interview might have seemed like a free-for-all because, by the end of their hour-and-a-half segment, they had shown off tattoos, danced with fans, received hugs and kisses, consoled crying fans, put long blue extensions in their hair (and in mine), and managed to get everyone to do the Mexican wave. I gave them the creative freedom to do their own thing live on-air because, regardless of their musical lives, they were just fun and decent people. And boy, did they exceed my journalistic expectations.

As the session came to an end the ladies performed two more hits, "Stop" and "Viva Forever." They were then whisked off to the airport. This interview was noteworthy for me because it gave me some primo cross-demographic recognition. My fan base had now expanded to include young female girls (and their mothers).

Ginuwine
(*Da Mix,* "In Da Round," February 8, 1997, April 7, 1999, and April 7, 2001)

As a VJ it's always awkward to focus in on light-hearted banter about music and beats when your subject has suffered a great loss. A few days prior to Ginuwine coming in to appear on a live "In Da Round" segment his father had committed suicide. Having lost my dad just

Much Master T

months prior, I could feel for him and I knew that the Ginuwine the viewers saw and the version I was interacting with were like night and day. No matter how hard he tried to mask his feelings, I could tell he was tortured by this turn of events in his life. He could have pulled a no-show and I wouldn't have sweated anything. I definitely would have taken some time off if I had been in his shoes. But people use whatever methods they've got at their disposal to get through shock and pain.

As I watched Ginuwine doing his sound check before the show, I really felt the need to pass along some kind words. I don't usually intrude into the personal lives of my guests, but I thought it was necessary to break this rule on this occasion. I told him that I was sorry to hear about what had happened with his pops and I recounted my own experience of losing my father. I told him that it's okay to take time to grieve, even though he was still in the early stages. Paula had given me some taut advice when I lost my dad, so I passed this along to him. As unbearable as things might be, she told me, don't run away from your feelings. Just try to move through the grieving period and feel the pain.

As Ginuwine told me, he had been forced to confront his feelings on the way to Toronto. He told me that, as he was sitting on the plane to Toronto, a flight attendant came by and asked if he wanted anything; he just cried uncontrollably. I reassured him that that it was okay. As cliché as it sounds, it takes a big man to cry, so I gave him a brotherly hug. It's rare, in the biz, when two people can take a moment like this away from the lights and the cameras. At that point in time, we were just two brothers, two human beings who were hurting. Still, it was sad to see him in this condition. I remember reading, a bit later on, that he had developed an alcohol problem and

Celebrity Interviews

that he was having a tough time confronting his own demons. But on this night, he channeled all of his energy into his performance.

When I saw him at a festival in the Bahamas in 2000, he was looking pretty worn and tired. Like earlier at Much, it appeared that he was using his performance as a coping mechanism: it was over the top. He was sweating profusely and grinding away at the speaker. It was kind of like R&B lover boy overkill.

At the time I still hadn't dealt with the loss of my father to the extent that I should have either. The potent combo of my father's death and Paula's post-partum situation almost took me out for the count. The way I was feeling was no surprise though; the specialist we talked to said that it was not uncommon for the woman to recover from these feelings and then have her partner fall apart shortly after, from trying to keep things together during the course of the situation.

This trip to the Bahamas was monumental not only because I got to experience the combination of grieving and warm weather first-hand, but also because some weird coincidences started cropping up. A trip to cover the Bahamas had come up many times before, and for some reason I had always passed it up. When I finally decided to take it, Paula felt queasy about it. She felt that something unusual or odd would happen on the trip and she couldn't explain why. Call it woman's intuition. And lo and behold, on the very last day of my stay in the Bahamas I had one of those life-altering moments that one could only chalk up to Godspeak.

All the work was done and out of the way, and I had the whole last day of the trip off. Sandra and the cameraman had left the day earlier, so I was on my own. I whipped out my book and CD player and was ready to take in some sunshine and get some well deserved

Much Master T

R&R. I sat on the beach and began reading Iyanla Vanzant's *Acts of Faith: Daily Meditations for People of Color* in preparation for an interview I had coming up with her in a few weeks. After a while I went out for a casual swim. I came back to the beach for a bit, then decided to head back into the water to walk around. I had heard that the year prior there had been a hurricane and that the bottom had not been really cleaned. Sure enough, as I was walking around, all of a sudden the bottom of my foot felt really strange. At first it didn't dawn on me what had happened, but then I noticed that a lot of blood was oozing out. Something in the water had seriously cut me under my big toe and made another small gash near the toe next to it. I panicked somewhat but tried to focus on this blood situation. I went back on the sandy beach and sat on my deck chair.

Why did this happen to me, I thought, and on one of my rare days off? Prior to this incident, I had cut a tendon in my finger and I had to wear a splint for about a month, then perform hours of rehab to get it back in working order. Whenever such a major event occurs, I wonder what the underlying message that I am supposed to be getting is. Admittedly, there were a lot of things that I did not want to address in my life, mostly having to do with discipline and order. The finger and foot injuries symbolized so many things that Paula was trying to get me to look at for too many years to mention.

I kept thinking about what Paula had sensed before I left for the Bahamas. She did not know exactly what would happen, but we are so connected that she could feel that it was something that would get all my attention. My mind was racing. I was also thinking about blood loss and about my fear of contracting some little-known tropical illness. I finally gathered up some courage, put on my sandals, and walked down to the waterfront to wash off the blood, hoping that

Celebrity Interviews

maybe the salt water would disinfect the cut. I asked the area groundskeeping crew workers to call someone to help me and I ended up being placed in a wheelchair, though I didn't need that. I needed to see a nurse, and fast. When I finally got to see one, she looked at my foot, bandaged it up, and said I had to go to the hospital. Then we got into the whole public versus private hospital debate. The debate boils down to this: for private care you have to shell out U.S. dollars. But hey, I'm a celeb. The nurse knew who I was, and so did her assistant.

They put 12 stitches in my big toe. It hurt like hell. When they were done, I hailed a cab back to the hotel and called Paula on my work cellphone; she was happy to hear that I was okay. I couldn't wear any shoes, so I was hip hopping around in the hotel lobby when something in my subconscious told me to turn around. I did—and I saw a large trail of blood behind me. I was weak, exhausted, dehydrated, but my foot was still bleeding, so I had to head back to the hospital. The doctor I saw (a different one than I saw earlier) told me that I had a bloodclot because one of the arteries hadn't been closed properly. So he had to undo the previous stitches and find the problem area, fix the artery, and then sew up the gash again. This doctor was a very caring soul who kept saying "yes, daddy" after every little procedure he did. He would be saying, "now I have to close up the artery, daddy, and it's looking good, yes, daddy, and just two more stitches, okay, daddy," and so on. At first I kept thinking to myself that he had to be joking, but after a while he put my mind at ease. Four hundred U.S. dollars later and with painkillers in tow, off I went by cab, again, to the hotel.

I was being wheeled through the hotel lobby this time around when I bumped into my friend Alton, a former Much cameraman. He

Much Master T

apparently had come down to Bahamas on vacation. Can you believe my ass was attempting to make plans with him to go out that night, because it was my last chance to party before I left the next morning? Then it suddenly dawned on me. What was I doing? I realized I had to slow things down a bit. Maybe I had been taking on too much (no pun intended) and maybe to some extent, like Ginuwine, I was running away from feeling all the losses. I wouldn't slow down and take stock, so the finger and toe slowed me down to allow me to start reassessing my life and spending time with myself.

By the time I had hooked up with Ginuwine for a third "In Da Round," he had changed his stage set-up to include a lot of background singers. One guy performer in particular appeared as if he had a dual purpose, which was singing and security. Again, *Da Mix* was not *Hard Copy*, but it was public knowledge that Ginuwine had contemplated suicide at one time or another. As a VJ I had to ask myself, how should I deal with this topic on a light prime-time video show? Certainly, talking about it with Ginuwine would allow us to illuminate the topic for kids who might be going through similar issues. So we broached the topic of his father's death again, and he looked hurt. But he did give us some lasting advice. He said he couldn't let his father's suicide destroy his life because he had way too much to live for. I don't know how our viewers felt, but these words of wisdom still ring true for me today. I think we all have too much to live for, if you want my views on suicide. Part of that, no doubt, was the fact that his woman, rapper Sole, was pregnant with his child at the time. I think that made him realize that he couldn't do the same thing to his kid that his dad had done to him as a young man. As somebody who had been left behind by a loved one, I think he had a whole different perspective on the issue.

Celebrity Interviews

Mary J. Blige

(*Chillin' with*, February 1997, *Da Mix*, May 19, 1997; *Chillin' with*, August 1999, *Intimate & Interactive*, October 20, 1999)

I can count on one hand the number of musicians I consider to be true friends in the industry, and the queen of hip hop soul is certainly one of them. My relationship with Mary started when I first interviewed her during her promotional run for *My Life* in 1997. Leading up to the interview, the clippings I read claimed that she was unprofessional, that she had a bad attitude, and that she was disrespectful to journos. I knew that there was no way she could have been this bad, so I wanted to use the interview as an opportunity to dig around for the hidden Mary the Good. The interview took place in the traditional five-star hotel. Sometimes these environs can make interviews quite sterile. Mary's make-up person was in attendance as well, and when Mary entered the room she commented on how much she liked my jewellery. Never mind Mary's music, her attire was something to behold as well. She sauntered into the hotel room wearing a flashy gold outfit that didn't suit her quiet and reserved personality at all. It was as if she was in a shell and ready to break out. This was all about self-discovery.

Before we went anywhere with this interview, I had to broach the topic of why the media had saddled her with this difficult-to-work-with label. It seems that a few years of getting ripped off by dubious labels and management, receiving short royalty, and getting stiffed by other shady characters in the industry had left quite an impression on her. This was the real reason that she had become such a

Much Master T

guarded artist. Straight up. After her *What's the 411* album, she didn't see much of the proceeds, and when you're putting yourself out there and not receiving your proper dues, it becomes discouraging. I'm well acquainted with that.

The thing I appreciate most about Mary is that she's not afraid to be herself; she's down-to-earth and she's the epitome of "real" People who have seen my interviews with her have told me that I was able to bring out a different side in her. Unfortunately, the footage of my first interview with Mary was shaky. Rebecca Rankin (who later moved to VH-1) was a new videographer at the time and she shot the interview. I didn't hold out much hope for salvaging footage because I was watching the camera tilt from side to side as I was asking Mary questions. But we made the best of it and put it together for a *Chillin' with Mary* show.

Karen Gordon, a renowned publicist (whom I first met while doing the *New Edition* interview) handled the next session I had with Mary in Toronto. Mary had come to town to do an "In Da Round" and, during the day, she did a segment for *The New Music* with Larissa Gulka. The word around Much was that Mary hadn't enjoyed her interview with Larissa too much. Apparently Larissa asked a question and Mary was evasive, so everything that had been said about Mary before was suddenly thrown at her again. In addition, Mary was pissed off because *Mic Check* magazine had published a story linking her to Dr. Dre, which was alleged to have been factually incorrect. Apparently she had called *Mic Check* to cuss them out just prior to her interview with Larissa. When you couple that with the fact that coming to the Much Environment during the day is like being in the pit at the TSE, you can see why Mary J. was on edge.

Despite her negative experience with *The New Music*, Mary still

Celebrity Interviews

showed up to guest-host *Rap City* that same afternoon. She gave me some kind words regarding my approach to interviews ("You make things so easy"), which made me feel all warm and toasty inside. Later, she showed up for her spot on "In Da Round." The buzz surrounding this show was unprecedented, with a huge number of people from all ages gathered outside the studio to take it in. We packed the place up to the rafters. When Mary started singing, the excitement reached a fever pitch and the energy in the room was electric. She was decked out in a white suit, her hair was dyed blonde; she was a true diva in her ghetto fabulous garb. I was equally ghetto–fabbed out (I was wearing this great tan suit from Hoax Couture with a nice matching yellow shirt). This segment makes my Top 10 live performances for too many reasons.

With the release of Mary's *Share My World* CD in 1998, I had the opportunity to travel to New York to interview her again. On my way to the Big Apple I listened to Mary's latest release. When I got to the hotel, we heard that she had been up all night shooting a video so they weren't sure whether the interview would take place as scheduled. It was looking like we'd have to stay overnight potentially. I was not comfortable with this because I had not stayed overnight in New York since my father had passed away. But I didn't want to let on to the Universal reps about my own personal issues. I was apprehensive and wanted to head back home. So we were all hanging out in the hotel room for a couple hours when Mary's sister, Latonya, came in and gave us the heads up, to alert us that Mary was still sleeping. I was just thinking about going out and picking up some PJs 'cause it looked like a sleepover. Much to our surprise, a half-hour later Mary came into the room looking like she had just woken up. She didn't have any make-up on and she told me that the only reason she

Much Master T

came out was because I was there. I was really appreciative and flattered. Mary was as comfortable in my presence as I was in hers.

The hotel room we shot the interview in was nicely lit. It was moody. It felt like an evening shoot. I'm not sure if it was because we caught her in the morning, but she let it all hang out. She talked about her relationship status, some personal problems, what she does to find happiness. She opened up her heart and soul to Canada. It got so good that she started goading me on to talk about tight-lipped stuff, like her relationship with K-Ci, the hurt she went through, and how she had to move on. Our viewers got a close-up of why Mary rules. It's because her music reflects what the average woman is going through. She lays it on the line and pulls no punches.

Part of her being so forthright during this interview also had to do with my charm and media savvy, I'd like to think. I'd interviewed her previously and TV doesn't lie. In fact, when I asked her what it was like working with Lauryn Hill, she said it reminded her of sitting with me — like a friend.

The next time I hooked up with Mary was at the 2000 Soul Train Awards. She had just won the Sammy Davis Jr. Award for Entertainer of the Year and her publicist was running around telling people that she wasn't granting any interviews. I thought that Mary would talk to me if she knew I was there, so I waited for her to break through the crowd. Once she saw me she was so ecstatic about winning the award that she fell right onto my arm and rested her head on my shoulder. I gave her a kiss on the forehead and that was it. It was a natural moment, like congratulating a good friend. She was fatigued and famished, so off she went.

In 2000, Mary came by the studio for an *I&I* and it was the most intense I'd seen her yet. The night before the show, the band did the

Celebrity Interviews

soundcheck and they wanted me to come in and meet her at night (I don't like meeting interview subjects beforehand, as you already know). Mary just happened to be sounding really intense and I can a tell mood swing when I see one. Her throat was wrapped up with a scarf and they were working on the background vocalists, working on harmonizing (apparently they had a last-minute fill-in vocalist).

The performance was the next day and the crowd was entirely comprised of Mary-philes. That level of fandom is so rare. Her outfit was ghetto fab to the nines, a yellow furry coat, denim trim, denim mini, and a white wedge boot that came up over the knee with straps in the back. Orange hair, big hoop earrings, and lots of finger jewellery. Problem was, Mary was bitching the night before and didn't want to do certain songs. She thought there were too many ballads in the set. During her performance she was apologetic about doing some of the slow songs. She began to remove certain songs from the set, and three-quarters of the way into the *I&I* she was finished. We still had to fill out the time because the show is planned right down to the last second. As Mary launched into the classic remake of "It's Not Over," Much staffers were freaking out because Mary had put together her own playlist right at the end of the show without consulting our staff. This type of thing would inevitably throw the production of a live show right off 'cause everything has to be timed perfectly.

Through the IFB in my ear Sheila Sullivan was saying, "She's not supposed to be doing this song right now, why is she doing this?" I was equally confused, but I wasn't surprised — this was classic Mary. Blige was ready to close up shop and that would have left us with 15 extra minutes to fill. A natural disaster. We could have thrown a couple of videos on, but that was not in the schedule at all. So during that commercial break, it was pandemonium running wild. Our floor director,

Much Master T

John Campilis, tried to sway her into performing another song. Mary was just not receptive to him. Her sister came over with her red leather pants in tow. And Mary was shaking her head, saying no. It took the president of Universal Records to finally convince her to go back on. I came out of the commercial break wondering what the heck I was gonna do. I was playing it off on the air like everything was fine, and I asked her if she was gonna do another song, but she still wasn't pleased with the song selection. The reason she wanted to re-jig her set was because she didn't want to bore her fans. She's not a bitch, she's smart and totally about the fans. She apologized that she had to perform another ballad, because she wanted to end with a fast song. Finally, the bandleader agreed to do a more up-tempo number, and she was relieved.

What you, the viewer, saw on-air was not what went down. We had to perform some creative editing to eliminate the drama. During the *I&I* she'd be asking aloud to the audience in a joking manner, "Do I look fat?" "Is my skirt twisted?" That's what you gotta love about Mary, she keeps it real. During our Q&A she was equally revealing, admitting that she looks up to Lauryn Hill for inspiration and that one day when she was down and depressed Hill gave her some kind words that carried her through. My last correspondence with Mary came when she sent a message to me for my Goodbye Blocko. George Stromboulopolous conducted the last interview with her and that was good — for now I had to let go of Mary.

Celebrity Interviews

Erykah Badu
(*Da Mix*, "In Da Round," March 1, 1997)

At Much we used to receive an average of about 10—20 new CDs per week. Most of these would end up in a slush pile, but others would grow on you. When I received this promo buzz single entitled "On and On" (with no accompanying artwork), I was hooked after my first listen. Once I found out the name of this R&B vocalist with the Billie Holliday flow, I kept my eyes and ears open for her full-length album. When I heard Badu was coming to Toronto on a promotional wave, I immediately decided that I would plug her into one of my "In Da Round" segments. It was a natural decision because, even though her album hadn't really begun to affect the public consciousness as yet, I sensed the simmer would quickly hit the boiling point as soon as people allowed themselves to make the musical shift. Not only did *Baduizm* alter the musical vibe, it was also responsible for initiating shifts in fashion, poetry, and urban culture.

Around this time Denise and I had a discussion about the quality of the "In Da Round" presentations that I was producing. One of her concerns was that the audience might be a little disappointed that their urban music heroes weren't getting the same treatment as the *Intimate & Interactive* artists when it came to dressing up the Much set, doing things with lighting, etc. The funny thing is that I started producing "In Da Round" out of necessity; I didn't feel that enough deserving black artists were getting the *I&I* treatment. I certainly understood Denise's concerns at the time, but it was more important that the fans were able to get up close and personal with their fave artists. Yes, it would be great to have the Much Environment dressed.

Much Master T

But the thought of not seeing Erykah Badu in the "In Da Round" format because of things like lighting and set decoration made no sense to me. Obviously, the problem was that *I&I* had a major budget and "In Da Round" small budgets, sometimes none at all.

In Erykah's case, we did manage to get a small budget happening to make the ambience more dramatic. I contacted her record label (Universal) and requested about $500 to get the set dressed. That's peanuts by industry standards, but Marlis Vos, the set designer, could work her magic on a minimal budget. She was brilliant at making the studio look like a million bucks on camera, and she had an eye for capturing an artist's aura when decorating the set. We wanted a space that reflected Erykah's spiritual essence and this is exactly what Marlis conveyed. Not that Badu needed any help. She was able to project a mood through her music and her stage presence alone. When she appeared on "In Da Round" she came out in a headwrap, she waxed on about African consciousness, and she burned incense on stage. The audience was mesmerized by her stage antics. How could you not love the fact that she wore Egyptian ankhs and fine silver without it letting her clothes steal the show?

At the time of this interview Paula was very pregnant. It was only a month before she gave birth to Kalif, so she watched the segment from home. That day, I was pacing around frantically. Some staffers probably thought I was walking around aimlessly, but it was with a purpose. I was just trying to collect my thoughts, tap into the energies of the room, check the audio, talk to the tech director, check out the crowd, see who was coming in, and think about Paula. It was always important for me to see the audience members (I refer to them as "soulmates") so I could make sure they were comfortable and that they weren't stuck out in the cold.

Celebrity Interviews

At the end of the show, I asked the soulmates if they had experienced something powerful and wonderful that night and they replied with an emphatic "yes." The reason I asked is that Badu had spoken about being a vehicle for the Most High. She said that her gift enables her to send messages through her music. She also spoke about change — how the universe was ready for change and how change was in the air. After the interview Badu went out to greet fans and sign autographs. Her words really made an impression on me. It was a tumultuous time.

Six months after that "In Da Round," I interviewed Erykah again at Kingswood Music Theatre. This time around Badu was pregnant with her first and only child, Seven. After the interview was completed I told her that my wife was in the audience with my newborn son and she asked me where they were hiding out. She said she'd love to meet them and asked if I could get them. I didn't think there was any way Paula would come up to meet Erykah (she never comes to meet any of the artists). Paula and I had just come back from Jamaica, and it was a turning point in Paula's life because she was going through her own stuff, though being in a different environment helped clear her mind and made her feel more confident as a new mother. So I went to get Paula. Her reply was, "For what?" After much cajoling (and to this day I'm still surprised she actually did it), she brought Kalif to the private room where Badu and company were staying. When we walked in, Erykah's eyes just lit up. She was looking at Paula's locks, which had reached her bum by now. Y'see, there was this big deal over what exactly lay beneath Badu's headwrap at the time, and the popular suspicion was that she was hiding dreadlocks. It was later discovered that she wore fake locks. Then she shaved her head bald.

Much Master T

She was really sweet to Paula. They talked about motherhood and when her baby was due. Paula told her how flawless and beautiful her skin was and I had to agree. She was absolutely glowing; motherhood certainly agreed with her. She asked Paula if she could hold Kalif. This is when things went wonky. She was sitting in a chair and she took the baby. She held Kalif with one hand supporting his neck and head, and the other supporting his bottom. Then she turned Kalif upside down six or seven times in a row in a quick rocking motion. Paula's eyes opened wide and she was trying to say something, but nothing came out. She gestured every time Erykah did it; she wanted to intervene but she was too freaked out to do or say anything. As Erykah's back-up singers casually looked on, Badu explained to Paula that this was an African ritual. The theory was that if she held Kalif close to her stomach immediately after she rocked him and her stomach moved in a certain way repelling Kalif's energy, that meant her baby was a boy. If it was a girl she said her stomach would be receptive and calm. All of a sudden she started murmuring, "I'm really feeling girl energy, it's definitely girl energy" and then she proceeded to hand Kalif back to me. Despite all this mysticism, she wound up having a boy. Still in disbelief over what had just transpired, Paula got another little shock when Badu gently kissed her hand on the way out. She had never had a female celebrity kiss her hand before and she was slightly caught off guard. Badu certainly left an indelible impression on my clan. Interestingly enough, the next time she came to Toronto she wasn't granting any interviews to anyone. Later, when Paula and I talked about the whole thing, Paula suspected that the reason Erykah really wanted to meet her was to flesh me out; to see if I was keeping it real. She wanted to know if my wife was truly a "sister."

You'd never know by looking at this photo that Jennifer Lopez had just split up with P. Diddy and I was dealing with my own issues. J-Lo hung in there for this *Live@Much* on February 13, 2001. (Photo by Barry Roden, courtesy MuchMusic)

This was my last *Intimate & Interactive* at MuchMusic. I couldn't have worked with a better artist to close that chapter of my life. Shaggy ripped it up that night and showed North America that reggae music was here to stay. (Photo by Barry Roden, courtesy MuchMusic)

The Spice Girls, spicing it up live on July 12, 1998, for the MuchMusic *Intimate & Interactive*. I loved working with these ladies 'cause they really knew how to work the media and never took themselves too seriously. (Photo by Mark O'Neill, courtesy MuchMusic)

This was the first time the Backstreet Boys came to the MuchMusic environment for their *Intimate & Interactive* on January 4, 1998. Hosting this event opened me up to a whole new audience, even though I was hesitant about doing it. (Photo by Barry Roden, courtesy MuchMusic)

It's nice when you get to work with an artist who is totally chilled, and as you can see for this installment of *Live@Much*, Janet Jackson and I were just hanging. (Photo by Barry Roden, courtesy MuchMusic)

I always find this picture funny. Much's official photographer, Barry Roden, kept asking me to lean closer to Madonna, so I kept leaning almost to the point of falling out of the chair. Madonna finally said to me, "You don't have to lean *that* much." (Photo by Barry Roden, courtesy MuchMusic)

Celebrity Interviews

The next few times I ran into Badu were at the Soul Train Awards. Erykah was hanging out in the arrivals area. She looked over at me, smiled, and then kissed me on the cheek. We engaged in some chit chat, and I told her that she seemed really free-spirited at her Toronto concert, which she did. "I am free," she told me. Because Badu and I vibed so well, she granted Much a pre–Soul Train Awards interview over *Entertainment Tonight*. Wow. I turned to the lady that worked for *ET* and she said facetiously that I'm gonna be a real problem, snagging all the major stars right out from under her nose. That was my coup d'etat.

Janet Jackson
(New York, September 3, 1997, and *Live@Much*, April 19, 2001)

I was nervous as hell when I was granted the right to interview one of the Jacksons. Who wouldn't be? Next to Michael (at least before he started having all these surgeries, making claims of "vitiligo," putting on surgical masks, buying Elephant Man bones, sleeping in oxygen chambers, etc.), Janet was the most admired Jackson. But I figured hanging out with Janet would be a little different than hanging out with The Gloved One himself. I should also make a confession: I've always had a major J.J. crush, even in her chubby days. I enjoy her any way she comes.

The reason for the interview is that Janet had just released a new album. *Velvet Rope* is such a sexual CD that you had to know the line of questioning couldn't help but be a little intimate. Janet had sold

Much Master T

over 50 million records worldwide and was putting her sexual self out there for all to see. She was no longer the cute, squeaky-clean kid on *Good Times*; she was acting more like an adult now, showing all of this skin — in the liner notes, in videos, everywhere. She had just made a guest appearance on David Letterman in which he seemed to be enjoying a lot more than the conversation. I guess he got roped in; I couldn't help myself.

We were flown to New York for the interview and put up in a $400-per-night hotel. It was the kind of place where even the burgers are high-end, ringing in at something like $20 U.S. a piece. We then were shown to a suite where we were to hang out and wait for our scheduled interviews. There were four or five journalists from New York and other parts of the States, but I believe I was the only one from Canada. Janet was running late. As we waited, some of the folks at Virgin were kind enough to show us some pictures from a photo session they had conducted with her.

She was looking as sexy and sultry as ever, and these photos were borderline soft porn. Janet was wearing a latex-type outfit and her nipples were noticeably pierced. Naturally, every male journalist was foaming at the mouth, just loving every minute of it. Incidentally, I was frothing at the mouth for other reasons. That tea and cookie I had earlier weren't nearly enough food to energize me, but I've made it a rule to never eat a large meal before interviews. I have this theory that eating pre-interview weighs down my energy.

When Janet was finally ready, the publicists brought me down to another room for the interview. I didn't travel with a camera person this time because these larger-than-life artists do their own make-up and take care of their own set-up.

The room where we were going to do the interview was a wide

Celebrity Interviews

open space, though not overly huge, with a couple of modern-looking chairs, as well as the usual camera and lighting elements. Janet was sitting in one of the chairs and greeted me with a handshake. She looked like a natural beauty. She wasn't wearing too much make-up, like most celebs, and her naturally curly, flowing hair looked lovely. I got seated and I was pretty fired up to begin the shoot. In cases like this you're on their schedule and they move the press in and out pretty quickly, so there was no time to waste. I remember that the room was rather hot that day, and I could certainly feel the energy Janet was putting out as we began. I wanted to capture the essence of Janet and her, ahem, sexual habits during this interview. Janet had opened the flood gates by discussing things like masturbation on this record, so I had to come out gunning and find out what this all meant from her perspective. A few questions in, I asked her about the different images she puts forward in her life: one very sweet, almost virtuous, and the other more hedonistic and sexual (like in the pictures the journalists were shown upstairs). In mid-sentence, she interrupted me and said, "Pictures, what pictures?" I was a bit freaked out when she told me to hold on one minute. Then she called somebody over to discuss this picture situation with him. Apparently, below the radar, her publicists had shown a bunch of mostly male journalists these images of Janet in the semi-buff, and she knew nothing about it. It seemed like one of those situations where record company heads were gonna roll. The word in the hotel suite was that no journalists were allowed to see any more pictures after that.

After the picture fiasco was dealt with, Janet was very open and candid in our interview. We talked a lot about how sexual the album was and I got her to admit that, yes, little Janet was having erotic

Much Master T

sexual experiences. She also gave the viewers some insight into her relationship with her then husband Rene Elizondo, talking about how much she leaned on him at times (to the point where she felt that she was overbearing). She explained that she would never have made it through some of the struggles in her personal development without the man in her life. I left the interview happy that it had gone even better than I had thought possible. Definitely an eight-and-a-half out of 10. When Janet Jackson appeared on *Live@Much* four years later, I was the man for the job once again. This time she was promoting her *All for You* album, which was dedicated to her fans. By the time she arrived for the interview, I had already decided on a few areas of focus for this tête-à-tête. I wanted to get her to discuss her body image issues because it seemed like her weight fluctuated quite often. Like a seasoned vet, I eased into this line of questioning by complimenting her on her abs and asking her what her workout routine is like. She responded by saying that her workout regimen is pretty hard to stick to, but that her personal trainer keeps her in check. At this point, it was time to go in for the VJ kill. I slyly men-tioned how society can be cruel to women who are public figures when their weight fluctuates, and that's when I got her to spill the beans. In fact, I was pretty bold on this one, asking her if food is a source of comfort when she's depressed. I remember her admitting that "I eat if I'm bummed out" and saying that weight issues are all subjective. "If you like a little bit more meat on your bones," she commented, "so be it. If that's where you feel good about yourself . . . It's all about yourself."

Another area I focused in on was her deeply private personal life. If there's anything you must know about the Jacksons, it's that they don't want anybody up in their business. However, deep down as a VJ, you know you have to go after a little dirt. I asked her, point blank,

Celebrity Interviews

about her now ex-husband, Elizondo, who was threatening to reveal her private life if he didn't get a good cash settlement in their divorce. She must have had to field this question more than a few times because she didn't flinch. She told me that she felt slightly betrayed, but that the Jackson juggernaut would go on. I further nudged her into discussing what it's like to lead such a private life, despite being a very public figure. Looking somewhat shaken, she replied that living a normal life is difficult and that it was unheard of, in her line of work, to have a relationship that lasted as long as her 13-year marriage to Rene. All of this candid relationship talk was the perfect segue into talking about sex. Her latest record had all of these implicit sexual innuendos and I just had to get the 411 on it. I told her, "People are surprised at how much of a sexual person you are," with impunity. She said she's always been seen as the squeaky-clean girl next door and that's why people have been making such a big deal out of her "new" sexual side. "Oh my God! My little sister is having sex now," she said.

Sadly, whenever fans can't pinpoint who's sleeping with whom, they tend to fabricate a bunch of rumours around issues of sexuality. There were more rumours like this flying around Hollywood than you could shake a stick at. According to rumour, both Janet and Michael were bisexual. I boldly quizzed Janet on this perception. She admitted that this wasn't the first time she had heard these rumours, and explained that hanging out with some of her dancers had caused the misconceptions to occur. "I'm very close with the dancers and a few of them are some of my best friends, so we're very affectionate people. Dancers, period, are very affectionate people. Someone comes into the room and you're sitting on their lap and they think, 'There's something going on,' but that's been with me forever."

Much Master T

By the time I hit my top five questions, things had already started to get a little hot. There was nothing I couldn't ask her. So I said, "Aside from personality, what attracts you to a man physically?" She came right back with, "Ooh . . . his package." I said, "His what?" and again she said, "His package." After this playful banter, I knew I had conducted a tight, solid interview.

I don't think any other VJ would have had the audacity to broach these subjects with Janet. Looking back, I'm glad I had the chance to interview her in April 2001. My days at Much were coming to an end, and it gave us a chance to pick up on the strong connection we had made in New York. In all my years on-air, I never put in requests to interview artists. But when Janet *Live@Much* came up, I knew I was the VJ for the job.

Backstreet Boys
(*Intimate & Interactive*, January 4, 1998)

When Denise Donlon first told me that she'd like me, Master T, to interview the Backstreet Boys for a special *Intimate & Interactive* show, I told her straight up that I couldn't do it. Her response was, "What do you mean you can't do it?" I told her that I had already planned my weekend. I was going away with my wife Paula. I stared directly into her eyes and said, "I'm going to Niagara Falls with my wifey and no child, hint, hint . . . can you visualize the 'Do Not Disturb' sign on the door?" Obviously Double D (this was my nickname for her) had double vision, 'cause she wasn't buying it. If there's one thing you need to know about Denise it's that she is

Celebrity Interviews

persistent. No, relentless. You don't get to be the president of Sony Music Canada without some kind of chutzpah. Wanting to know what it would take to abandon my plans for our romantic rendezvous, she continued badgering me to do the show.

I was feeling this tremendous burden to conduct the interview for Much. It certainly didn't help that I didn't have much of a desire to interview these guys. I know it was unprofessional, but the main thing underlying my reluctance to do this interview was the fact that I just didn't like the music. At the time, the Boys had sold 20 million records worldwide, and two million of those in Canada alone. The refrain, "Get down / Get down / And move it all around" had become a part of our public consciousness.

To make matters worse, this was one of those interviews that took Much a long time to confirm. The details of the interview were firmed up on the Friday and the interview was slated to happen that same Sunday. Donlon played hardball, telling me that she'd be willing to have a car pick me up from Niagara Falls and then take me back to continue our getaway. The toughest part of this ordeal was that I had to tell Paula that our mini-vacation/freak session was being cut in half because of Backstreet mania. It's funny, but she saw this as a golden opportunity to expose myself to a whole other audience. Ultimately, I caved in to the pressure and asked Denise to get me a car to drive me to Toronto from Kitchener.

The weekend played out like this: Friday afternoon, I drove from Toronto to Kitchener and dropped my son off at his Nana's. We then zoomed off to Niagara Falls where we spent two nights of fun and frolic. Early Sunday morning we hit the road back to Kitchener, where we hung out with Kalif for a short while and I began preparing myself for the *I&I*. I was just about to knock back a lovely cup of tea when

Much Master T

my mom informed me that the airport-style limo was ready to go. (Could you imagine, after all that, they didn't even send a stretch! Who cares, my mom was the talk of the neighbourhood.)

On the way to Toronto I was really feeling important, but I soon had to snap out of that headspace and get down to the business of fine-tuning my questions for the Boys. I had done all the research during my down time in Niagara Falls (thank God Paula finally went to sleep, if you know what I mean). When I arrived at MuchMusic I was blown away by the magnitude of the Backstreet phenomenon. The fans, the volume of their presence, and the heightened atmosphere all contributed to the hysteria. I slid through the back door of the Much building and headed straight to the Much Environment. Denise greeted me with a sigh of relief and we set about the task at hand. In a typical situation as the *I&I* host, one usually goes through a day and a half of intense rehearsals. But because this show was locked in only a couple days prior to airing, all rehearsals happened that afternoon. I remember watching the group do a soundcheck; it seemed so rehearsed and choreographed. It really lacked vitality. It seemed like they were holding back. Maybe it's because they had performed a show the night before to a crowd of 27,000 at SkyDome. In any case, when I looked outside the Much building a few hours before the set-up, the crowd absolutely blew me away. Queen and John was filled to capacity. I convinced myself that maybe this could be a fun experience after all. The fans were mostly young girls and they had been lining up outside the Much building since 1 a.m., crushing themselves up against the barricades outside. The cult of celebrity thing was in full swing.

Much had to open the parking lot out back to deal with such large numbers (people were guesstimating just over 2,500). This was an

Celebrity Interviews

I&I, but in no way was it intimate. Large numbers of these youth had to stand outside in the freezing rain to catch a glimmer of their heroes. There were reports of sprained ankles and hypothermia and Much had to set up an emergency response centre. The police presence was crazy, as nearly two dozen police officers and private security kept the audience at bay.

The director of this particular *I&I*, Dave Russell, decided that it would be amazing to intro the show with a shot of me in front of the building, walking and talking to the fans. I had to hype the crowd up, welcome the TV audience, and formally introduce the Backstreet Boys. The music kicked in and the group dramatically launched into a rendition of "Get Down," at which point everybody moved it all around. After they performed a song off their new album entitled "Hey Mr. DJ," they started to get the audience involved in the proceedings. Asking one side of the crowd to say "Hey Mr. DJ," while the other side sings "Jam All Night Long," they got the audience to see who could shout it the loudest. At this point it was pure pandemonium, dozens of girls were screaming at the top of their lungs.

Again, when I introduced them individually it was deafening. The *I&I* was in full motion. Having interviewed as many bands as I had — everybody from Boyz II Men to 112 — I had learned to look for the one guy in the group that you can connect with. For me, Howie D (Howard Dwaine Dorough) was that guy. He seemed to be the voice of reason amongst the five. Kevin, the eldest, was probably the most forthright throughout the interview. AJ was pretty quiet except for when he was complaining that it was too cold to go outside to the window to greet the die-hards (sorry, but there was no way he was going to get out of this one). When we returned back inside the studio the Boys performed an a cappella version of "Quit Playin'

Much Master T

Games (With My Heart)." It was then that I realized who the real main vocalist was in the group: Brian had some pipes and he could actually sing (pre-fab concerns notwithstanding).

The interactive banter, as usual, didn't cease to amaze me. A young girl named Christie from Manitoba asked the group what they'd be doing if they weren't Backstreet Boys and Howie said he'd be "wishing he was a Backstreet Boy." They also had to answer some pretty private questions. One caller asked whether they sleep in boxers or briefs, which made them kinda squeamish. If anybody cares, Nick was the only brief-ophile. During the interview they also got into how they put together their dance routines. These boys revealed that they were choreographed by L.A.-based Fatima Robinson. They might as well have been choreographed by the Jacksons or New Edition. The whole appropriation thing was a little difficult to swallow.

The Backstreet Boys closed off the show with a live performance of "Everybody." At the end of the day, the *Intimate & Interactive* had gone well. Shortly after this interview I started to notice something about the type of response I was getting from fans at the street level. My audience had grown to include a new demographic. Not only was it all about the cool, urban hipster crowd, but just about everybody and their momma was included in the mix now. This new addition to my audience was a loyal and passionate bunch. Just think about all of those teens standing outside in the freezing rain to watch the show through the windows. The whole thing made me think that maybe the fame clock on this group wasn't set to 14:59 after all. My ears are still ringing from the roar of the crowd that day.

After the interview, Denise Donlon was ecstatic. When I got home I got to watch the rebroadcast, the show seemed effortless. Denise was a pretty wise exec because she always told me that as a vj on

Celebrity Interviews

these *Intimate & Interactive*–styled shows I had to be just a traffic cop directing e-mail questions, faxes, audience questions, and phone calls and fusing them together with my own questions.

Nailing the interview with the Backstreet Boys was a major accomplishment for MuchMusic. It set off a trend that made it much easier for us to land interviews with high-calibre, multi-platinum artists. I've got to admit that this is one of those times when one has to give props where props are due. If it hadn't been for Denise's foresight in going after these interviews and pushing me to rise to the occasion, Much wouldn't have grown into the worldwide mega machine it is today. In retrospect, I'm really glad I did this interview because it allowed people to see that I could interview anybody. After the Backstreet Boys, a whole string of segments aimed at a crossover audience opened up for me. And ran with it!

Madonna
(*Live@Much*, March 6, 1998)

As MuchMusic prepared to host the Madonna *Live@Much*, the station still hadn't decided on a VJ to conduct the interview and host the show. So, how did I land the gig? A senior rep at Warner Music Canada later told me that he was the one who pushed for me to conduct the interview, because he was impressed by the *Intimate & Interactive* that I had done with with Jewel. I allowed her to feel comfortable and relaxed, and things flowed smoothly. Even I was surprised by how well it had gone. More importantly, the Warner rep also told me that I was one of the few remaining Much VJs who had the ability to put my ego aside.

Much Master T

When Much finally granted me the interview I wasn't as jittery and excited as everyone expected me to be. I had a lot of things on my plate that week and honestly, I just didn't have the time to get nervous until a few major things were out of the way. However, since interviewing Madonna fell slightly outside of my realm of expertise, I knew that I had to do three things to prepare: research, research, and more research. It was the first time in her career that Madonna had ever done an interactive format in which she'd have to field questions from her fans in person, via phone, fax, and e-mail, so I knew it would be a learning experience for the both of us. What made things more complicated from my perspective is that Genevieve Borne of MusiquePlus was also slated to host certain segments of the show. For some reason, the thought of having a co-host seemed to ignite some of my insecurities.

As the show approached, it became very clear to me that, if I messed up this exclusive Canadian TV interview with the Material Girl, my career could end up in an immaterial world. And this pressure wasn't coming from the head honchos at Much. It was being applied by my fellow VJs (who also doubled as Madonna groupies) and by the legions of streetcorner music critics who were tuned in to the goings on inside CHUM-City building. Of course, I was also putting a little pressure on myself.

Regardless of the methods I used to escape the pre-Madonna interview hype, I just couldn't get away from the buzz. Complete strangers and Much staffers kept asking me whether or not I was nervous, wondering what I was going to wear, and bugging me about what I was going to ask her. I really had no answers for any of them. None. During prep sessions for the Madonna interview, all the rules seemed to change. As far as I'm concerned, journalists are

Celebrity Interviews

supposed to keep a safe distance from their subjects so they can ask questions with some kind of objectivity. But, oh no. Not this time around. Every-body wanted me to treat Madonna as if she were a living goddess.

The week of the interview, Denise still wasn't convinced I had this Material Girl thing down. I had been concentrating on the K-Ci and Jo-Jo "In Da Round" scheduled on *Da Mix* for Wednesday, and Madonna was to follow on the Friday. I remember Denise asking me if I had watched all the interviews that MuchMusic had done on Madonna. Denise's mantra was to watch as many tapes of past interviews as possible in order to be fully prepared. This time around, she was worried for a myriad of reasons. Clearly, her reputation as Director of Music Programming was riding on it. As you know, I was always well prepared for interviews, but I liked to keep it loose and spontaneous whenever possible. Somehow, I don't think that's what Denise wanted to hear. Honestly, I didn't feel the least bit intimidated by Madonna, which is the key to doing an interview with her. If she senses that you're in total awe of her, she'll have you for lunch. It's all about timing, really. You have to know when to ask the scandalous, uncomfortable questions without getting her in a David Letterman–like tizzy — Letterman once slyly suggested that she was a serial philanderer and the Queen of Pop blew her stack on-air and vowed never to return on his show, although she did years later. For me, this interview was an opportunity to show the Much brass that I could reach a wider, more mainstream audience as a VJ, despite the fact that I'm not the biggest rock or pop junkie in the world.

Regardless of what you think about Madonna, there are at least two things you have to respect her for. First, she's not afraid to take a chance on something she believes in. And second, she's truly bril-

Much Master T

liant when it comes to marketing and constantly re-inventing herself artistically. Plus, she's filthy rich and powerful (I guess that's three things). At the same time, she's been criticized for being what hip hoppers would call a "biter." This means that she's made a career out of appropriating the musical styles and dress codes of different subcultures. The first thing that comes to mind is when she took the Vogue dance phenom (that I had witnessed years earlier at the Twilight Zone after hours club in Toronto) to dizzying commercial heights. I had to remind people that this dance craze had been around for years before she recorded her hit single "Vogue." It's kinda like she lifts trends from less visible communities and puts her own little spin on them.

The day before the interview Denise Donlon asked me, once again, if I had seen all the interview material with Madonna and I told her that I hadn't. I think she must have been concerned with my general attitude going into this whole thing. I think my naturally confident, carefree demeanour made her feel I wasn't taking this challenge seriously enough, though I was. When I got home that night I listened to Madonna's forthcoming *Ray of Light* album. I turned out the lights and really focused in on what she was trying to say lyrically. I think it was her most spiritual effort to date and I actually connected with the record. As a VJ there's nothing worse than not being super-familiar with an artist's work. Except really disliking it, that is.

When the day of the interview finally came, the police had to close down one lane of Queen Street and put up barricades to control the hundreds of screaming fans that had been waiting outside the Much building for Madonna since the night before. Before these big gigs are announced, the head honchos at Much meet with security personnel, paramedics, the local public transit authorities, and traffic control to

Celebrity Interviews

make sure things will remain under control. Security for the event was extra tight. It had been decided that Madonna was not going to go to the window to mingle with the fans, as most *Live@Much* guests used to do. The clamps were also shut down where other media outlets were concerned. *Hard Copy* and *Entertainment Tonight* had requested media access for this one, but Madonna's management had prohibited U.S. print media and photographers from getting anywhere near the star. Only Barry Roden, who had been hired by Much, was allowed to shoot Madonna. And even then, he was only allowed to do so for 15 minutes. As far as the lighting requirements were concerned, a one-page rider specified a 10K light and front-on shots only. There were to be no profiles.

In the early part of the day, Denise, Sheila (the supervising producer), Genevieve (a MusiquePlus VJ), and I had a brief meeting to go over our questions and map out our particular responsibilities for the show. This is when I learned that I would be seated with Madonna for the entire show and Genevieve would be moving around fielding questions from the audience. When the meeting was over I felt a sense of relief. Finally, I had a decent understanding as to how things were gonna flow. I had never really understood how they were attempting to weave Genevieve into the mix, but now that things were clear, I was able to focus in on my subject. I had visualized how this would all play out in my head many times before, and I knew what I had to do to make that visualization into a reality.

From there Genevieve and I went to the Much Environment chatting about all the excitement to come that evening. That's when I learned from her that she was a huge fan, which was perfect because I wasn't necessarily one. I like Madonna's earlier stuff and have loads of respect for her, but I'm not really into the pop thing.

Much Master T

Believe me, in this particular case, I think it played in my favour that I was more impartial. It was mid-afternoon when we got into the studio for a cue to cue (which is the technical term for a full rehearsal with camera, lighting, audio, and everybody involved). This would be the first time that MuchMusic and MusiquePlus would broadcast the same program live simultaneously. After the cue to cue, things were getting pretty heady for me, and I knew it was time to be introspective and get my thoughts together.

Exactly one hour before the interview I started jotting down my questions on cue cards. I always like to outline which questions go with which segments. I managed to write down three segments' worth of questions before I stopped. I got distracted by other things and I never went back to finish the questions, which was okay with me. I knew I'd eventually want to lose my safety net as we got to the end of the interview anyway. I had been in the boardroom for about an hour when I heard a knock at the door. It was Johnny, the floor director for the event. He insisted that I go down to check the lighting for the show one more time. As you can probably understand, this really irritated me. Why did they have to interfere with my groove and disrupt my calm at this time? To make things worse, when I finally did make it down to check the lighting one more time, the security guard at the main entrance to the Much Environment refused to let me in. He said I didn't have the proper wristband. "Come on, man," I said. "I work here. I'm *doing* the interview, remember?" I know security was supposed to be tight, but this was demoralizing beyond belief. Here I was, prepping for the biggest interview of my career to date, and this security guard (who had seen me around the building, by the way) decided to exercise some muscle. Finally, Sheila, my biggest supporter, came over and sorted

Celebrity Interviews

things out.

I went to sit in the chair from which I was going to conduct the interview. The crew were checking the lighting and I was bitching and overreacting to this security guard because he really pissed me off. A part of me more or less attributed it to my nerves, but Sheila was definitely getting an earful. Just short of having this guy fired, I reluctantly started to cool down and try to let people do their jobs. Sheila kept repeating softly, "It's okay, T, my big star, I know, I hear you, I hear you." When the necessary checks were finished, I went back up to the boardroom to compose myself before the interview. I was still a little shaken up by the episode with the security guard, so I knew I had to calm my nerves before the big event. Right there, in the middle of the boardroom, I got down on my knees and whispered a long passionate prayer asking for the Lord's guidance. Then I picked up the phone and called Paula. She dropped me some inspirational words of wisdom and I was ready for the Material Girl.

I scooted down to make-up quickly and was sitting in the chair when all of a sudden the make-up person attending to me got up and ran out of the room. She had been waiting all day just to get a glimpse of Madonna, and she had heard some commotion outside in the hallway. It was indeed Madonna's entourage hustling into the Much Environs. When my make-up person returned, she explained that she had seen Madonna. Apparently Madonna had smiled at her and seemed like a nice person. "T, you're gonna have a good time!" I was told. With my make-up complete, I quickly headed over to the studio. Marlis had done an impeccable job of dressing the set. It was soothing: just what I needed at the time. I intentionally looked over to the interview area where the loveseat and chair had been arranged for the show—and there she was, the queen of pop, with

Much Master T

her $10,000-dollar-a-day make-up person doing his final touches on her. Her eyes locked with mine. She smiled and I smiled back. From that moment I knew it was on.

It was an emotional voyage leading up to and during the interview because I didn't know whether to mention my 11-month-old child and the death of my father. But after I told Madonna about these things on the air, we seemed to develop a connection. I understood the deeper meaning behind the song "Little Star" (which is dedicated to her daughter) and the following cut, "Mother's Girl," which, as she told our audience, had been written in memory of her deceased mother. That was our little bond. After that, things seemed to go smoothly. You might even say that Madonna had me eating out of the palm of her hand. She convinced me to try Hot Tamales, one of the bowls full of candy that had been brought in especially for her (Madonna is a candy addict). Then she showed me a yoga move.

At one point after the first commercial break, Sheila said to me through the ear monitor, "T, it's going well, you're doing great," and this whole feeling of warmth and reassurance coursed through my veins. Getting that support from your peers is beyond important and for that I'll be forever grateful. Obviously I'm the man up front, the one that everyone sees conducting the interview. But a show on this level could never come off as well as this one did if we all hadn't been a team that day. Just knowing that Denise believed I could do it, Genevieve was holding it down in the audience, the crew was behind me, the production staff was in my corner, the fans were loving it, and someone like Sheila was in my ear passing on these kind words really had a calming effect on me.

When it was all over, I just stood there on the set. My socks were still off after the yoga routine and I just stood there frozen for five

Celebrity Interviews

minutes as the fans filed out of the studio. The interview had gone well. At the reception a bit later on, Madonna told some of the Much brass that it was one of the best live interviews she had ever done. What an accomplishment; I was relieved that it was finally over. When I bumped into Denise in the hallway afterwards she wasted no time in letting me know how proud she was, kissing me square on the lips. I was a bit shocked, but I was glad that she was so ecstatic.

Britney Spears
(January 29, 1999)

I interviewed Britney Spears just two months before her debut album, *Baby One More Time*, broke big, eventually going on to sell more than 13 million copies. Canadian kids were certainly aware of her, but I don't think anyone expected this Mickey Mouse Club alumna to become pop music's next sweetheart sex symbol. After doing my homework I was impressed to find out that she had been in the entertainment game for a long time, plugging away until her big break, and her staying power impressed me. But there was one small problem I had with Miss Britney: every time I watched her videos I couldn't help but think to myself, "Oops, they've done it again."

On the afternoon of the interview there were a bunch of kids standing out in front of the MuchMusic building. A producer and a security guard went outside to hand-pick some fans to bring them in to witness the interview. A few employees from CHUM-City also brought their kids in so they could hang out and meet Britney. I was introduced to her right there in the studio. When I first saw her,

Much Master T

against the backdrop of all those kids, I thought she looked like an average teen. She was wearing a blue shirt and a pair of jeans; she was a cute kid. As we got into the interview, I asked her some pretty generic questions like, "What was it like putting together your first full-length album?" From her responses, it felt like I was interviewing somebody who had been so pre-prepared for this press routine that she wasn't able to answer outside of the box.

As we were hanging out and vibing, I told her that I was the worst person to teach dance steps to. I'm incredible when it comes to freestyle, but when it comes to technically learning a dance step, this brother has two left feet. Hearing that, Britney decided to break down a dance called the Will Smith for me to try. Before I let her go ahead, I asked all the kids in the audience to stand up and give it a whirl with me. If I was gonna look bad in front of the nation I wasn't going to do it alone. As soon as everyone was set, Britney got into place and executed the move perfectly, providing a demonstration for the rest of us. Then she watched as we tried to replicate the Will Smith on the air. It was good fun, and by the end of it we all looked like a choreographed formation in one of her videos.

Just as Britney was set to come on the show, the issue of the former Mouseketeer's alleged breast implants started to become big talk at Much. People were swapping pictures on the 'Net and checking out magazines for comparison. For me, the whole thing wasn't even an issue. And that's a problem. Have the times changed so drastically that things considered taboo in my day have now become so commonplace that they don't even warrant a response? Couldn't something like this be damaging to an entire generation, what with so many young fans looking to their idols for guidance? Ultimately, it's her body and she's free to do what she likes. Either

Celebrity Interviews

way, there was not gonna be any boob talk on the show that day. Things were shut down, probably by her stylist, who fitted her with an oversized shirt for the walk-on session.

I interviewed Spears when she still had that youthful enthusiasm thing happening. Even then, she was all business and so little pleasure. So I felt a little used and abused in our interview. I think success does strange things to kids; it can't help but make a powerful impression on somebody so young.

Eminem
(April 29, 1999, July 4, 2000, and May 29, 2001)

How do I feel about white artists performing black music? Will the increased involvement of white rappers in hip hop mean the ultimate death of the music as we know it today? These are sensitive topics that provoke heated discussion. Certainly, some of my Caucasian brethren display a degree of talent, and Eminem is foremost among them. Company Flow's El-P and rappers like 3rd Bass have both skills and a respect for the culture that the art form emanates from. But others like Vanilla Ice have neither.

Inevitably, these debates pop up during any interview with Eminem. The first time I interviewed Em he had just come out with his debut record and he was in town to do a promotional gig at the Opera House. In spite of the media attention that he was getting, he really wasn't a big deal at the time. Did I come with my heavy-handed questions about cultural authenticity? Nah, I didn't need to. He wasn't supposed to blow up like he did. Then something happened. *The*

Much Master T

Marshall Mathers LP debuted at number one, according to SoundScan. A white rapper flipping rhymes about rape, "faggots," and "bitches" sold 1.7 million copies in a single week. Now, I can tolerate some of the ignorance in rap (and other music), but all of this violent talk troubled me. Point blank, I asked him about his lyrics. Why did he rap about things like putting his wife in the trunk and killing her and all that? He was sincere in his response. He just asked me if I had ever been mad enough to kill somebody. I told him that it was certainly possible, adding that I'd never write or rap about it for *Much* or MTV. He defended himself by claiming that what appears in his music is nothing more than his true, pure thoughts. As for the issue of white rappers in hip hop, I decided to let him off the hook.

My first tryst with Slim Shady was no life-altering interview (if anything, it seemed like he was trying to get respect from me as opposed to the other way around), but the second time around was different. By this time in his career, Eminem had become larger than life. He was performing as part of Dr. Dre's Up in Smoke tour, and I was granted the only one-on-one TV interview. It was a big deal at the time. Despite the fact that his second record was the subject of much controversy as a result of his homophobic lyrics, he had been anointed by the press (including *The Source*, the Bible of hip hop culture) as the Next Big Thing.

Denise Donlon asked me if I was going to question him about the homophobic content in his lyrics. I knew it was unavoidable. He had to be challenged. And it would be interesting to see how he responded to the charges. When you go into a high-profile interview like this, you can't just launch into scandal and controversy. You have to approach things more delicately or your subject is likely to exit the room faster than you can say homophobe. That's why I was careful

Celebrity Interviews

not to treat him like the monster that the media was making him out to be. When I finally asked him about the violence and homophobia in his music, he came back with an interesting answer. He said he was in a Catch-22: if he didn't continue to rap about hardcore issues in an explicit style, the hardcore hip hop heads would think he'd gone soft, perhaps pandering to the pressure of the media.

The last time I hooked up with Eminem it was during an interview with his homies, D12. By this time, Liz Doyle of Universal made me realize that there was a demand for my interview services, so I was sent to interview Em for the record label. Because we had built up a good rapport, I was the only person the rap star had requested for a Canadian interview. Universal knew that Em was a ticking time bomb because before I conducted the interview they had me sign a waiver release form; it stipulated that Much could air the segment only for a certain period of time. I had never before or after signed such an agreement, but that time I felt I had to, since Liz Doyle told me that Eminem could refuse to grant me the interview if I did not agree. The powers that be at Much were pissed off that I signed it, but we ended up getting some great material.

Having flown to Detroit on a small, private plane, we then drove to Birmingham, Michigan. From the jump I knew this interview with Eminem and D12 was going to be more challenging than a one-on-one session. When artists are by themselves they're generally more chilled out. But when they're with their boys, they feel testy and try to punk off. There were five of them and only one of me, so it would have been easy for them to pull the inside joke routine or just ham it up for the full interview. I knew I had to be aware of this going in.

As the interview got under way, it became clear that Eminem was carrying these guys on his coattails. They had been tight for years and

Much Master T

they had made a pact that if one of them were to make it, he'd have to come back and pick the others up, so that's what Em did. With this in mind, I wanted to discuss some race issues. If Em hadn't been white, would he have been able to pull these characters into the limelight?

The focal point for this interview was supposed to be D12, but it wound up being Em. In one of those incidents that producers dream about, Eminem absolutely blew his stack when I asked him about a track on his latest CD in which he disses Limp Bizkit's Fred Durst. As it turns out, the two rappers were supposed to have recorded a track together dissing Everlast. But Durst backed out at the last minute, taking shots at Em on MTV instead. Em responded by calling him a "pussy" on record. When I got him talking about this feud he just lost it. He was dissing Durst and swearing his head off for like three minutes straight. At this point, the interview had nothing to do with D12 anymore. Em had become the focal point. Again. He was possessed.

After the interview, Eminem went to hang out with his daughter and I went to the recording studio with the D12 crew. It was a diss session at its best. There were snaps, mama jokes, papa jokes, even T jokes. I held my own until one of the brothas said my locks looked like Kit Kat bars. Dem's fighting words. Good for him I had to take that private plane back to Toronto.

P. Diddy
(*Live@Much*, July 31, 1999)

You can call him Puff Crappy, Fluff Daddy, P-Diddly, or even Puff Snappy, but one thing you can't call him is a financial failure. No

Celebrity Interviews

matter what you say about Puffy, there are three things you gotta give him credit for: he is a true survivor, a brilliant businessman, and you can't stop him musically.

I've had the opportunity to interview him a couple times. The first time I interviewed him was in '98 during the Caribana festivities. He was in Toronto to promote the *No Way Out* record. I remember that the responses he gave to my questions about his jetsetting lifestyle (he had flown in on a private jet and his label, BMG, had to foot the bill) were distant and empty. He was acting as if he was on some promo schedule from hell. It seemed pretty clear that he didn't want to be in Toronto, but when I saw him again on a float for Caribana, it seemed like he was whining his troubles away.

As I was trying to enjoy the carnival festivities, I started thinking: how did P. Diddy manage to become a part of our everyday lives, melding in there right next to Mcdonald's, Jennifer Lopez, and Instant Messenger? The Puff phenom is just that, phenomenal. You gotta hand it to a guy who has built an empire on sampling old hits by artists like David Bowie and The Police.

Whenever Puff comes to town it creates major drama for the reps at BMG who have to meet all his demands. This time, they were reported to have included requests for a certain thread count mandatory for his bedsheets. So it really came as no shock that, as I was conducting the interview, I felt like I was talking to a brother who had insulated himself from the rest of the world through his wealth. He had trouble reacting to my questions. His responses were rehearsed; his answers calculated. Normally, I like to think that I've gotten to know my subject a little bit better after a couple interviews and that the viewers have been able to share in this experience, but that just wasn't happening here. Puffy always seemed to have his guard up, and it was difficult for me to pull anything emotional out of him. Could he possibly be devoid of a personality or was he able to exercise that much control in the public eye? When the interview was over, I remember mumbling to myself, this brotha is going through a lot of stuff.

By the last time we hooked up, the media once again thought that Puffy's star was already fading fast. I was also on my way out at Much, but I managed to book him for a *Chillin' with Puffy* session. Finally, in my fourth interview with the star, I was able to break through some of his protective shell. This was the most candid interview I'd ever done with him. I asked him if he thought he was invincible, and he looked at me and smiled, as if to say, "Boy, if only you knew." He made it quite clear that the support of friends and family kept him going during his trial situation, and I could see that it was the catalyst for changing his perspective on life. I continued to chip away, breaking down the thick layers that had, for so long, separated him from his fans. I asked him about the J. Lo situation and he looked genuinely hurt. It was refreshing to see that he had actually taken stock of some emotional issues in his world. When the interview was over, I finally felt as though I could actually sit down and relate to him as a person. And that's exactly what we did upstairs on the patio of the Nation's Music Station that day.

Dr. Dre

(in New York, November 3, 1999, and *Up in Smoke* tour, July 4, 2000)

Sitting alongside DJ Premier and Pete Rock, Dr. Dre is the best producer in hip hop, bar none. He was the mastermind behind NWA

Celebrity Interviews

and he made Eminem. So you can imagine how tense and anxious I was as I prepared Much's *Chillin' with Dre* segment. Along with people from *Peace* magazine (one of the leading print vehicles for urban culture in Canada), Universal Records flew me into New York to do the interview.

I was in the hotel room waiting for Dre to arrive when he walked in with his manager. He took a seat and Basil immediately hooked up a mic to him. Then out of nowhere Dre abruptly asked to be excused. Basil proceeded to perform the arduous task of unhooking the mic. Confused and surprised, I tried to figure out what just happened. What could have been so pressing? We waited for half an hour and finally he returned, apologizing. There was a certain sense of relief to his disposition, as if a load had been taken off his shoulders. That's when he dropped it on us: he desperately had to use the washroom. He said he preferred to use the one in his hotel room because he didn't want to kill us by blowing the place up. Instantly, Basil and I were in hysterics. Dre had set the tone for the whole interview.

I wanted this interview to show Dre as both a multi-talented producer and a hip hop icon, so I knew I'd have to hit him with some questions that he might not have been comfortable talking about. More so than usual, this interview had to be about tact. Dre has a very dramatic history with the media (he once slapped a woman, Dee Barnes, during an interview) and I didn't want to get beaten down or disrespected.

Coming into the session, Dre was concerned about how his new album would be received. His seminal album, *The Chronic*, had already been hailed as one of the greatest rap albums of all time, so he was feeling the pressure to duplicate this success. Still, he said he does his best work when the critics are on his back. Turning to

Much Master T

more controversial issues, I asked him a question about Death Row Records. He had had a falling out with Death Row's thug CEO Suge Knight after Tupac was killed, so I knew it would be a touchy subject. His answers were short and to the point. He said he harbours no ill will towards Suge. Having worked with Tupac during his final days, he mentioned how brilliant he thought the late Shakur was. He told us this great story of how he had laid down a rhythm track for Pac, and how the talented emcee and thespian had written a perfect song to accompany his beat within minutes. Dre also had plenty of Eminem tales to share, stories illustrating how proficient the Great White Hype is as a writer and an industry pro. Dre told us that he and Em would go into the studio at 6 p.m. and still be in bed by 11 p.m. — and Eminem would have laid down four songs in this time. Maybe the greatest story he told that day was of how he discovered Em. Dre said he heard one of Eminem's highly regarded underground tapes, enjoyed it, and pursued him for a label deal. Apparently, Dre didn't even know Em was a white guy until he actually met him. That speaks volumes about how talent is not marked by colour, creed, or class.

Of all the interviews I had read with Dre, nobody had ever asked him about his talents as a video director. Having screened the bulk of his videos on Much, I consider him to be a video vanguard as well as an ace producer (did you know he's directed videos for the likes of NWA?). Dre was impressed by my thorough research in this area, and because I had deviated from the usual line of questioning, we were able to delve into other sides of his character. As the interview progressed, I could sense Dre becoming looser, more casual, relaxed, and interested in answering questions.

When you interview artists of Dre's calibre, you can always tell if they respect your command of the craft by the comfort level of your

Celebrity Interviews

next interview. When I next hooked up with Dre, at the *Up in Smoke* tour featuring heavy-hitters like Ice Cube, Snoop, Xzibit, and Warren G, it was clear that he was relaxed and willing to talk — in spite of the time constraints. We opened up with the issue of hip hop and ageism. He was 40 at the time and he hinted, off the record, that this might be his last tour. This was a more mature Dre. He was in the game to take care of his family and make some money, rather than trying to create conflict (as he had done in his younger days). Still, we waxed about the prospects of putting together a new NWA record (which is still nowhere to be seen as of 2002) and we talked about the fact that he had a mobile studio with him so he could work on this record while on the road (where are the tapes, man?). So maybe he's not through with conflict and controversy after all.

R. Kelly
(*Da Mix*, September 2, 1998; Chicago, July 17, 2000)

Whenever I hook up with R. Kelly I call it the emotional meeting of the minds. The first time I hooked up with this multi-platinum selling soulster was in Detroit. Along with a number of other journalists, I was flown there to attend his concert before the interviews were scheduled. The show was full of energy and excitement. I hate to generalize, but Americans take in concerts on a whole other level than Canadians; they completely cut loose. Women must have sensed that R. Kelly had been locked in the studio for months on end 'cause they were losing their minds. The concert spanned the entire emotional spectrum. On songs like "You Remind Me" he

Much Master T

would objectify women (as his female dancers grabbed themselves and moved provocatively), and then he'd turn right around and perform a song that pays homage to his mother. A huge picture of his mom even dropped from the ceiling at the back of the stage as he poured his heart out, crying on bended knee to his mother. It was so R. Kelly.

The interview that followed the concert was part of the effort to promote his newly released album, *Twelve Play*; the beautiful Aaliyah just happened to be present in the room. For this session, the label only granted me the opportunity to ask him three questions. We were talking about the album's provocative lyrics and I asked him why this record was so sexual. He responded by saying that he was locked away in the studio for six or seven months and he was horny, straight up horny. We touched on some R. Kelly and Public Announcement (his first band), the early days, and we discussed the success of the group that brought him to this new phase in his career. My final question was about Aaliyah. I asked him what it was about her that made her so special. "Just look at her," he said, "she has that glow." He explained that when he decides to work with somebody they have to possess that certain light, that glow from within. She was sitting directly behind me, across the room, and I turned around and glanced at her. He was so right.

The question is: was he dating her at the time and did he marry her despite her legal status as a minor (she was only 15 and he was 27)? We now know the truth, but at the time of this interview, your guess was as good as mine. She would be giggling at times, looking over at him during the interview, and to me their chemistry was undeniable.

The next time I interviewed him I was able to get much closer to

Celebrity Interviews

his core. I jetted to the Windy City with Andy, my cameraman, to do a *Chillin' with R. Kelly*. I had gone over my questions with Paula the evening prior and I was prepped to capture something extraordinary for all his fans. I had a sense that this was going to be an incredible interview; don't ask me why, I just had a hunch. I was checked in to my room at the hotel and the interview was scheduled for 9 p.m. As I waited around, I was researching, re-reading an article on R. Kelly in *XXL* magazine, and listening to his double album again. It was clear from the *XXL* interview that he had already divulged revealing personal details in print, but I wanted to get him to elaborate on some of this stuff for the camera. His record rep called my room to say that the interview wouldn't be taking place as scheduled and that they'd keep me posted. I kept getting different hourly updates from Kelly's people, so by this time I had taken off some of my interview outfit so I could relax. By now I was exhausted. That king size bed was calling my name.

Around 11:30 p.m. I started to wonder if I was going to get this interview at all, and if I did, would I be in the physical and mental condition to pull it off? Finally, the call came through; I touched base with Andy, and away we went to Kelly's hotel. When we arrived we were escorted to an immaculate penthouse suite. A baby grand piano sat in the living room, which overlooked Chicago's impressive skyline. Andy began setting up for the interview and I instructed him to set up by the couch and light the piano area as well. Another TV crew from England came in shortly after us. I greeted them and made some small talk with Kelly's manager.

Then R. Kelly strolled through the door. He said what's up to his crew, greeted me, and apologized for being so late. He said he was playing basketball. Basketball? So much for priorities. He requested

Much Master T

a cup of coffee, took a few sips, and he was good to go. With the cameras rolling we started bantering back and forth about the fact that he had just finished playing basketball. That's why he was sporting his casual look for the interview, he explained. He was wearing a Green Bay Packers football T-shirt, a pair of white kicks, and diamond studs in both ears. As I said, we have a good rapport, very natural. It's almost like there's an understanding between us. So I asked him about the Public Announcement days. Immediately, he asked me to stop the camera. Did I say something that didn't jive with him? This was a first in my career. I had never had an artist stop the camera on me before. He proceeded to tell me that he can't talk about Public Announcement because the situation was going through the courts. Then he respectfully gave me the okay to continue and we resumed the interview. Whew! Close one.

When I asked him why his songs seem to resonate with women he told me he has a natural ability to feel what a woman's feeling, and to express it as a man. Soon after, we made our move to the piano, where he floated into a beautiful ballad about love and that special woman in a man's life. It was as if time stood still. Everyone stopped breathing, the room stopped moving. After he performed this amazing masterpiece, I noticed that his eyes had started to well up with tears. Sitting to his right, just two arm lengths away, I was truly captivated by his genius. After being swept away by his sheer musicality, I popped a question related to something I had read in *XXL* about how he likes to sleep in the closet (literally). "That's heavy," he said, admitting that, yes, he does retreat to the closet sometimes to take a break from humanity. Nobody asks him anything in the closet, he explained. Although he's a multi-millionaire with legions of fans, it was easy to feel for the guy. You could see

Celebrity Interviews

how the burden of celebrity was weighing on him to a certain extent.

The music industry is so small and insular that word of how much Kelly enjoyed this interview quickly got back to his people. As a result, they requested that Much send me out to cover the video shoot for the song "I Wish." At the time, there was a new show on Much called *The Making Of* that was designed to provide viewers with a behind-the-scenes snapshot of a video production in the works. We'd get full access on-set and it would be a half-hour segment. Again, the video took me to Chicago. They were shooting in the 'hood where R. Kelly grew up. Actually, the majority of the video was shot inside a cemetery, which was very emotional for me because I hadn't been in a cemetery since my father had passed away. It was a scorching hot day in the Windy City, but we had no trouble getting the overzealous kids on the street to give us their views on R. Kelly. He was a real hero, we discovered; a father figure in these parts. He was the clichéd boy-done-good. He had made it out of the 'hood and returned to shoot a video with his peeps. When I asked the kids to name their favorite R. Kelly song they said "I Believe I Can Fly." Right after that, they voluntarily began singing it aloud.

When R. Kelly finally granted us the interview, we set up on this little stone bridge overlooking a quiet pond. The sun was beaming down on the cemetery; reportedly it was one of the hottest days of the year. I was going through this process with thoughts of my father's death, and all the emotions that go with those thoughts, raging through me. A crowd of folks quickly assembled around us, which, for me, meant it was performance time. I had to turn it on — even if that did mean pushing thoughts of my father aside. Suppressing emotions usually means dealing with pent-up pain later on. But as my father taught me, the show must go on. I only had a 20-minute segment with R. and I made the most of it. I asked him what it was like to come back here after having lost his mother and many of his friends, and I shared with him my own thoughts about my dad. He said it was necessary to confront the past in order to aid the healing process. We talked briefly about the *TP2.com* album, and when the interview was through, his manager looked at me, beaming from ear to ear. This was one of those interviews that brought out many sides of me as a man. My dedication to my craft was not compromised when I had to deal with my emotions. By simply letting down my guard and just sharing I learned that my journey is a parallel path that's traveled by all, even the greatest stars. And I'm man enough to admit that this was truly an emotional meeting of the minds.

Jill Scott

(*Da Mix*, "In Da Round" September 30, 2000, and August 27, 2000)

When I received an advance copy of *Who Is Jill Scott*, I really was wondering who she was. As Paula and I headed out to pick Kalif up from daycare shortly thereafter, I popped the CD into the player and prepared for the Jill Scott experience. Whenever we listen to something for the first time, our mandate is to keep an open mind, regardless of what we may have heard about the artist or their music in the past. In this particular case, Paula and I were both pleasantly surprised. Paula was really blown away by the musicianship, and I was feeling the lyrics and vocal styling. The album was fresh and it had depth. There weren't any studio outtakes or fast-forward material

Celebrity Interviews

— an oddity in new millennium R&B. And, she wasn't trying to mimic anybody else's vocal style; she had her own distinct sound going on. I had the opportunity to check her out live at *Word* magazine's Urban Music Festival and, as one of the hosts that night, I introduced her by saying, "You're about to see an artist that's gonna make a strong impact on the music industry." I didn't realize she'd make these words a reality.

I remember having a conversation with Donna and Phil, co-proprietors of *Word* and organizers of this event. They told me that one of the reasons they booked Jill is that she had just launched her album. At this stage in her career, she was looking for an opportunity to be heard, so she welcomed a performance like this. Boy, did she ever make an impression. Jill's voice rang out like a bell in the air. Some of the best concerts Paula and I have ever been to are the ones in which you never have to look up at the stage because what you're feeling is so overwhelming that you get lost in the music and just take it all in. This was one of those times. Sandwiched between sets by Mos Def and Royce Da 5'9", Jill had her work cut out, but she pulled the audience in with her infectious energy. They were grooving with her every step of the way.

By the time Scott arrived on *Da Mix*, I had a dilemma on my hands. Should I let her perform a live set? How about a straight interview? Do I want her to do an "In Da Round"? Is she popular enough to merit an "In Da Round"? In reality, the decision was easy. I wanted to do an "In Da Round," but I didn't have a lot of pre-promotional time 'cause I had just come back from the St. Kitts Music Festival. This didn't give us much of a chance to build a large studio audience. To make things worse, my associate producer wasn't around that week to help us put the whole thing together, and the

Much Master T

reps at Sony were telling us that we had to do it now because she was going on tour for the next few months. I was on my own.

When it came time for the show, I was a bit disappointed. Without sufficient time to do promo, I settled for an "In Da Round" with no studio audience. I was planning a casual interview that would be interjected with performances. Because I was short a producer, things weren't quite so easy. Just like in my *X-Tendamix* days, I was running around trying to do a bunch of different things at once. Jill Scott was leery of performing three songs because her second back-up singer wasn't present. She wasn't a hundred per cent comfortable with that. This also made the process of choosing which songs she would perform difficult, and the indecision was eating up precious time when it came down to the sound check. Scott's keyboardist/arranger was getting a bit testy because of this indecision, and he couldn't figure out who was in charge of the show. He saw me as only the host, so he didn't think to direct his questions and concerns my way. It was a juggling act carrying this one off. Normally, I was involved in overseeing the live production part of the show (in addition to acting as the host), but I usually relied on an associate producer to perform specific tasks such as putting the rundown into the computer and performing a bevy of organizational duties. I know there were times when I was asking Jill questions and I was stuttering slightly; my concentration was elsewhere.

Jill Scott's bright aura and powerful presence carried me through this interview. I was taken aback by her charisma. She has a regal way about her, though she's still very much down-to-earth. Unintentionally and subliminally, she calmed me down. The last song she performed was "A Long Walk," and it was one of those rare musical treats where you could feel the energy throughout the televi-

Celebrity Interviews

sion studio. I was glued to the monitor, just as Paula and Kalif were transfixed as they watched from home. Whenever you can get a three-year-old to sit still and take in a performance (without promising a treat for a job well done), you know it's something special. Nothing Jill Scott went on to accomplish was surprising after this day. I wasn't shocked when she was nominated for a Grammy; nor when she received a Soul Train award; nor when her album went double platinum; and especially not when she released a critically acclaimed live album. We helped to break a new artist in this instance, and I'm proud of that. Sacrifices have to be made for excellence.

Sade
(New York, October 18, 2000)

If there's one female artist who gives this red-blooded VJ the willies, it's Sade. She represents all that's wonderful about a black woman. She's classy, charming, and she has the best smoky contralto in the biz. Honestly, I had waited almost my entire career to interview her and I wasn't sure if I'd ever see that day. Paula and I were both diehard fans, so you can imagine how excited I was when I found out that I'd have the opportunity to sit down with her for *Da Mix*. Originally, she was slated to appear live on the show, but a death in her family forced us to put things off. Three months later, I flew to New York to chat with Sade at Sony Studios.

This was one of the few times in which I found myself in an awkward position over the course of my long career on the air. I was a huge fan of my subject and her work; as a VJ I always tried not to

Much Master T

blur those lines. But I've done everything to Sade's music. I've cried to her music, made love to her music, taken long drives with my family to her music, and even at our wedding, "Your Love Is King" was the song we chose for our first dance. Paula and I once went through a spell in the mid '80s when our TV broke and we listened to Sade non-stop. Even today, hearing her debut album puts us right back in touch with our early twentysomething reality. To make things even more complex, it was starting to look like my days at MuchMusic might be numbered. Given the personal part that Sade has played in my life, I felt like my interview with her might help me to begin the process of ending my 17-year journey.

After years of waiting, the time had finally come. I boarded the plane and arrived in New York an hour and a half later. I checked into this happening hotel on the Lower East side. Then I began the process of making sure my questions, my outfit, and especially my emotions were in check. Before I left Toronto I went to Hardcore Designs and they hooked me up with a custom-made outfit (pleather pants and a matching wine-coloured satin shirt with an oversized pleather collar). As my Dad would say, "That's a bad hook up!" Just before I left the hotel, I turned on BET. I was getting the jitters for obvious reasons. After I went over the questions I had prepped for my Sade interview, I called Paula in Toronto. Unfortunately, we got into this argument that still hadn't been resolved when I jetted. The conversation snowballed into this big, ugly verbal exchange. It was massive, so huge that neither of us can remember what we were fighting about. Whenever I'm prepping for a large world exclusive interview, Paula's usually by my side. Emotional support from your best friend/significant other is critical as a VJ. It's not that Paula didn't care about how this Sade interview went, more than anyone

Celebrity Interviews

else she knew how important and special it was, not only for me, but for both of us. But I had taken her to a bad place and there was no turning back.

Although Paula was steamed at me, she later told me that she was praying I'd be able to channel the anger I was feeling towards her into something good. I had waited too long to take a crack at Sade. In fact, Sade, Burning Spear, and Oprah were the only three artists atop my VJ interview wish list, and we both knew it wouldn't be long before I would proudly hang up my clip-on mic. As I was trying to hail a cab in Manhattan, I was feeling lousy. I was close to tears. So I called Paula on my cell and pleaded with her to help me come to some sort of resolution. I couldn't go into the interview like this; I was desperate and off centre. I was trying to make her see that nailing the Sade interview was something special for both of us. She was grappling with my words and I wasn't certain if I could regain her support. As I neared my destination, I knew we had to wrap things up. Knowing that she wouldn't hear from me again before the interview, she said that she loved me and we resolved to work things out later. She also reassured me that I would do an incredible job and I told her I loved her too. We said goodbye and I felt a lot better. I was ready to redirect my energy into making this dream a reality.

When I finally arrived at the Sony Studios, I was greeted by Sade's publicist, who was British, which immediately put me at ease. When I finally got the call to go in and meet Sade, the production crew fitted me with a mic and then I sat down to compose myself for the interview. The set was well-lit and there was a warm, pleasant feeling circulating in the room. I could tell that the crew had been enjoying Sade's presence all day. The vibe was right on. The moment Sade arrived and said "hi," I turned to mush. She was more than 40

Much Master T

at the time, but she had aged nicely. She had cute little freckles on her face and we made great eye contact. She was wearing a long, flared jean skirt with a deep red, fitted blouse that was unbuttoned to expose a black bra. "T, focus, focus," I told myself. When I saw her and I realized that we were both dressed in red tones, it made me feel like we were on the same page for the interview. Her first words to me were, "Haven't I met you before?" and I nervously replied that I had met her eight years ago. I relayed the story of how we had met when she was being interviewed by Michael Williams back in the day.

The first thing you notice when you speak to Sade is that her voice has plenty of bass to it, a low-end with a slightly gritty edge. Even though she sings low, I expected her voice to be smoother and a little higher. She had been absent from the music game for eight years, a long hiatus spent raising her daughter, and she had taken some time for introspection. This new album, *Lover's Rock*, was her most contemplative to date. But I wanted to have her take us through a little career history before delving into the hiatus and the new album. Throughout the interview I could tell that she was impressed by my line of questioning. I was more than prepared for this gig 'cause I had absorbed a lot of info over years spent following her career. I brought out a few things about Pride (the original band she hooked up with) and their intricate connection, and I got some insight into her backup singer, Leroy Osbourne and their amazing relationship on and off the stage. She really got into reminiscing about all the memories that went into developing their original sound and forming the band. We also touched on some of the new material and what motherhood means to her. Even though her record label was pleading with her to come back years ago and do another

Celebrity Interviews

album, she stood her ground and took the time she needed to be away from the biz. She wanted to put her energy into living the kind of natural existence that she feels most comfortable with. I could totally respect all that.

People who saw the show said that I looked a bit like a salaried groupie, all gushing and gooey. Maybe Sade sensed it too and could have hammered me, but she figured that I'd come to interview her with good intentions. After a while I gave up struggling to fight the fact that I was smitten with her. When I left the interview she did the European greeting thing, giving me a kiss on both cheeks. Yeah, I had scored. When I got into the cab I realized I could do anything now, even leave MuchMusic. I had reached my career peak and I felt empowered. What a natural high. I took a deep breath and called Paula. It was a dream fulfilled for both of us. It was time for a new day.

Jennifer Lopez
(*Live@Much* February 13, 2001)

She's been called a sexpot, a space cadet, and a love goddess. I'm still not sure which is true, but it's very rare to interview someone considered by both the media and public to be one of the most beautiful women in the world. This interview took place at the height of her career; her sophomore album, *J. Lo*, had debuted at number one in both Canada and the U.S., and *The Wedding Planner* was number one at the box office. She already had the multi-platinum *On the 6* to her credit and had starred in several other hit films. When you add

Much Master T

her much publicized romance with the Artist Formerly Known as Puff Daddy to the fuss over the dress she wore to the 2000 Grammy Awards, you can understand the hype surrounding her *Live@Much* appearance. There would most certainly be a boisterous crowd at the corner of Queen and John.

This career-boosting interview could just as easily not have happened, because it got postponed twice in favour of American promo shows. The first time I heard that I'd be interviewing J. Lo, I was really excited, but when it was re-scheduled, I lost some of my zip. It just wasn't there. At the time, I believe, I was questioning myself, especially my skills. Through this cloudiness, I simply lost my focus. For a *Live@Much* event of this magnitude, my to-do list was always in check, but on this occasion I just couldn't get everything together. The afternoon before the big show I was really feeling stressed out, so I decided to take a nap. This is definitely not a regular practice of mine when preparing for larger productions because it takes so long to defunk myself after some deep sleep. I was just not myself.

One might think that I was putting a lot of pressure on myself for this interview, because J. Lo was the bomb. But this had nothing to do with J. Lo. At the time, I was going through contract negotiations and I felt like I needed to nail this interview to maintain the respect of both my peers and management. I felt like I had to prove myself again. My insecurities were trying to get the better of me. I had been on the air at Much for 11 years and I desperately needed to feel like I still had the juice. A few days before the *Live@Much* with J. Lo, I was on my way to Belleville for an annual festival. I had the J. Lo press file from Much and I was going over a few press clippings, but I was having a hard time concentrating. I didn't know which direction I

Celebrity Interviews

wanted to take the interview. Once the record company told me that I couldn't ask about her relationship with Puffy — one of the main reasons, aside from her obvious physical assets, why so many would be tuning in to check out Miss Lopez — I sensed that the record industry filters would kill some of the spontaneity.

I was in a pre-production meeting the day before Jennifer came in, and my producers were talking about having a Jennifer Lopez look-alike contest. I didn't think this was gonna work. The contestants were to perform a J. Lo dance or song, and I thought it was really lame. To this day, I don't think Much took J. Lo seriously as an A-list artist, and that's why this cornball idea came up. I don't think we would ever ask Bono to participate in something as cheesy as this. Paula helped get me ready by hooking up an appropriate outfit for the show — something that was loose and casual to relieve some of my stress.

I remember arriving at the Much building in a rush 'cause I was late. This *Live@Much* was to begin at 6 p.m. and I had to be ready for 5:30 to get things positioned for the opening. This particular show was unusual because the regular Much Environment was being renovated. J. Lo's interview would therefore take place in the smaller studio where EC and *Breakfast Television* are shot. Naturally, both myself and the crew were not accustomed to working out of this spot and there were some challenges. For me, this proved to be another unnerving distraction. I rushed down to the studio and it was five minutes to air time. On top of everything, we didn't get to do a rehearsal before going to air. I was notorious for bitching when asked to continually rehearse for these things, but this time we all really needed it. Talk about flying by the seat of your pants. Oh, I forgot to mention that I hadn't yet laid eyes on Miss Lopez. The crowd waiting

Much Master T

on Queen Street was really rowdy, especially the guys; the ages ranged from early teens to late 30s. The amount of testosterone was unusual for this kind of event. I'd guesstimate that a few hundred fans were outside the studio. It was a good crowd, but not as many people as O-Town and the Backstreet Boys drew a few weeks prior.

The countdown to air began and the audience started screaming and howling. I took a deep breath, still knowing I wasn't really on my game; deep down inside I wanted to make this one count. I opened with a solid introduction and everyone slowly settled down to prepare to welcome Jennifer to the Much Environment. She made her way into the studio towards me and the room was at a fever pitch when Jennifer and I made acquaintances. This was the very first time I had ever laid eyes on her and I thought to myself that, up close, she's not just an attractive woman, she's really hot. Her midriff was nicely toned and her army fatigue–style shirt and hip-hugging gold pants caused me and everybody else to do a double take. I greeted her, we sat down, and I got into the interview process. It was quite evident to me that she had matured into something other than what I recalled on *In Living Color*. Back then she was more heavyset and had more of an underground street vibe. It's funny, she was hot and all, but immediately when we locked eyes I could tell that I didn't have as much of a connection with her as with other artists I've interviewed. There was a chilly vibe — maybe even a cold vibe.

We got off to an energetic start. I asked a few questions, but the commotion created by the fans was distracting. The audience was really revved up and asked some very taboo, honest, and provocative questions. The audience always has the freedom to ask tough questions, because they don't have producers or multi-national record companies breathing down their necks. If a call, e-mail, or fax

Celebrity Interviews

comes in from a fan asking what Jennifer's proportions are, there aren't any professional repercussions. We took a call from somebody named Nick who asked, "What do you find most attractive in a man, and do you prefer Latino guys?" In true A-list celeb syle, she did a fabulous job of skirting around this issue. She laughed heartily and replied, "I don't know . . . You don't have to be perfect-looking to be sexy. I like that more. More attractive that way. I like Latin guys. Absolutely." Obviously, J. Lo didn't want to disrespect her Latin male demographic by making some off-the-cuff remark about not enjoying Latin men as much. Her taste in men is definitely across the board. One of the most touching things J. Lo said in this interview was that "Romance is not so much the expensive things — it's the little things. It's the notes. It's the call in the middle of the day when you're really stressed out. That kind of stuff is important to me. It's the everyday, day-to-day stuff that makes a relationship worth it."

The audience interaction during this interview reveals why the *I&I* segments are nothing short of revolutionary. Could you imagine being able to sit right across from your idol and ask a heartfelt question? This is what happened repeatedly during the show. Another teen e-mailer asked her a timely, topical question about body image issues, explaining that she had weight problems and that she'd "die to have a body like yours." Jennifer responded by relaying her own tales of battling with her weight. "I also fought with my weight when I was growing up," she said, referring to her days as a dancer on *In Living Colour*. "Those were my chunkier days. I'm not ashamed of my chunkier days." She also gave out some good advice on diet and exercise: "I just watch what I eat and I exercise. I'm like anybody else, I'm human. Sometimes I feel like working out, sometimes I don't."

Much Master T

The studio of the CHUM-City building, where the crazed fans converge, was open for the beginning of the interview. Normally we keep the window closed during the interview and open it only in preparation for the artist to go out and greet the fans. Immediately after that it's supposed to be closed again. But for some reason the window remained open for the entire interview. And of all the shows for this to happen with. Lopez' male fans abused this privilege by shouting out all kinds of things, running the spectrum from pleasantries to vulgarities. It was very, very distracting. It got to the point where I would ask her a question and she would be interrupted by these obnoxious male groupies (yes, they do exist, contrary to what most people think), making it difficult for her to respond. One loud and boisterous male fan in particular had her uncomfortably laughing throughout the interview; he was interrupting her every response. Good fans are hard to come by, but this guy had no studio etiquette whatsoever. Half-jokingly, even J. Lo admitted on-air that "He's distracting!"

It's not easy to carry on an interview in which your subject is constantly losing her focus. But once things piped down a little, I went back to the romance questions and asked her to tell us about the most expensive date she'd ever been on. She said it was "probably on a boat" rented for one week. Wow, a week-long cruise is one marathon, pocket-breaking date. We could only speculate that her suitor was Puffy. But I couldn't ask her about it. Everybody in the room wanted to know if J. Lo was still dating Puffy on the downlow. But there was no way I could ask her about it. I got the impression that her heart was heavy at the time; if she could have blown off the interview and not done it, she probably would have. The typical *Live@Much* set-up (where fans are within earshot of the talent)

Celebrity Interviews

doesn't work for everybody. Lopez seamed taken aback by the close proximity of the audience; she seemed overwhelmed by their heightened fervour.

She had just finished doing an hour long in-depth interview with Jana Lynn White at MuchMoreMusic prior to my segment. In retrospect, I really wish I had been given the opportunity to have a cool, relaxed sit down with her. I would have been able to get a lot more out of her and possibly some insight into everything she must have been going through at the height of her career. That would have taken some work though 'cause she's very private about family and very guarded with her emotions. I presumed that all the events happening in her life at the time (in addition to the audience's boisterousness) set off her mood. Her answers seemed very short and to the point, almost as if I was slightly annoying her at times. She exuded a certain nippiness and I didn't feel a compassionate vibe from her at all, a rarity considering the relaxed, genteel interview style I was known for. She was spelling out predictable answers-by-the-numbers, which was mind numbing.

The other thing that was driving me nuts was that we kept taking her to the window. By the end of the interview she asked, "Are we going to the window, again?" I felt her pain. But what with all the interest in J. Lo's booty, I used these trips to the window as my opportunity to get out the magnifying glass and inspect. Hell, I was curious, too. So I tried to be casual about it, sneaking a peek. Let's just say I don't understand what all the fuss was about. Her glutes were grand, but not huge by black standards. When I escorted J. Lo to the window for a third time, somebody suddenly yelled out, "Jennifer, you've got the biggest ass." I turned around and looked at this person like I was going to chew them to bits. The place went

Much Master T

awkwardly silent. My first instinct was to go over, because I spotted this girl in a red sweater, and just show her out. But we went to the window and came back and security had removed her by then. So not only did you have all of these testosterone-filled male J. Lo groupies hooping it up outside, but you also had aggressive young women on the floor yelling out their cheap one-liners, too. Man, was that distasteful. It was one of those bizarre incidents that periodicals like *Frank* jump all over. *Toronto Life* even ran a brief blurb on this incident, "because it was something right out of a B-movie."

The J. Lo look-alike contest we had planned went off just prior to this mind numbing incident. We had selected five Lopez look-alikes and we asked them to either sing or dance like Lopez. At this point, we were having major audio glitches that flattened the segment even more. This is where that no rehearsal thing really reared its ugly head. I wasn't really sure what was happening for this segment. Talk about getting caught with your pants down. It was embarrassing to have these young girls (some paralyzed in J. Lo's presence) attempting to do their thing to white noise. Even Lopez complained saying "It's terrible you guys don't have music for them!" At this point I got desperate and asked her to dance and show them something, anything, but she shot down my request. "I'm here to judge today," she said, sounding a bit put off. The whole thing was a bit of a disaster. The music didn't cue properly and the contestants weren't sure what they were supposed to be doing. I kept telling them that this is their golden opportunity to dance, sing, anything; I found myself scrambling to save the situation. It was a very painful TV moment for me.

Still, Jennifer was very gracious. She knew the segment was weak. I did, however, manage to pull off one major coup. Without going against the label's request that we refrain from asking her

Celebrity Interviews

anything about Puffy, I did manage to get into the issue. "When your heart is broken," I asked her, "what do you do?" She said that when she's finished with someone the party is over. No hanging around as friends, that's it, she's done. "It's hard. It's bad. I'm the type that just walks away — later! It's not that easy, it just has to be done," she said. The Much staff was happy that *Access Hollywood* ended up using this clip, but the interview, on a whole, was pretty lame. Funnily enough, I can't count how many fans told me how much they loved it that night. Yeah, right. If you say so.

When the ordeal was over I just walked around in a daze. Normally, post-interview, my energy is high, everyone who worked on the program is usually feeling really pumped, and they converge on me for some post-interview banter. But I knew deep down that this segment just didn't click. I was really hard on myself, but it took me a hot minute to realize that not every interview is gonna be great, much less good. Normally the next couple of days on the street you get the audience reaction, and the response was good, but I think people were just starstruck by the fact that I got to sit down and interview J. Lo. The average viewer doesn't focus in on the eye contact, every question, the flow; they just get caught up in the moment. What's obvious to me, after so many years in the business, is just not so clear to the average viewer.

After the interview a co-worker persuaded me to go back to Bravo where Much staffers had set up a little green room for J. Lo and her entourage. Usually I'm not into this whole rub, but for some reason I went along on this occasion; perhaps I was hoping for some acknowledgment. I think it really says a lot about my spirits at that time. Jennifer was there with Benny Medina, who used to be Puffy's manager. I looked at her and she seemed mopey and sombre. I was

Much Master T

gonna say goodbye as she left, but I left it alone. On her way out, she and Benny walked past me and he said a quick goodbye and she said nothing. I didn't know how to take it, but one thing was clear: she was emotionally spent. Just days after the interview it was announced that Puffy and J. Lo had definitely split.

Paula and I had a long talk after the interview at home and, as usual, my Mom called to say that she really enjoyed the show. That's when I started to realize that it's so easy to get consumed by the Much machine and start believing that you are what you do for a living. Maybe I believed that my job defined who I was, and this was a huge wake-up call. I wasn't guaranteed this job, as I found out during contract negotiations. Even though I was still doing it at this point, nights like this weren't magical and I wasn't on par. It was a humbling experience to realize that I am just me, Tony Young — not Master T, VJ and producer. As I came to realize this, I accepted the fact that the road out of Much would be a rocky and emotional ride. Change was in the air for me once again, and my J. Lo moment was bittersweet, at best.

India Arie
(April 21, 2001)

When Flow 93.5 FM launched in February 2001, I would drive around in my jeep listening to this song with a catchy hook by a young woman with an unconventional R&B name. As I listened to her sing about women not needing to subscribe to fake, unrealistic ideals of beauty, I also realized that the song had unusual lyrics. The message

Celebrity Interviews

resonated with me, and the track became one of my personal faves. This is how black radio in Canada impacts the general marketplace — by breaking songs that nobody else cares about.

As I heard this song being killed on Flow, I was putting together a show on sexual imagery in urban videos. I had programmed a show around this same topic in 2000 and I wanted to do a follow-up segment. At this time, the standards for rap videos — in terms of the sexual imagery — were hitting an all-time low. For the most part, rap videos had devolved into scenes featuring a bunch of scantily clad women dancing to the delight of the male emcee. It was (and is) sexist, misogynist, and tasteless, at best.

India's song inspired me to jump all over this topic. She was one of the few artists coming out and saying that appearing half-nude in videos does not define womanhood. There are other facets of femininity out there. Her song made an impact on me and I knew I wanted to have India on my show from the first time I had heard it. That doesn't happen often. And I'm male. So you can imagine the impact the song might have on girls and women. It's empowering and uplifting.

Pulling this show together was very stressful because it was only Petal and I, and we had put together a panel discussion and lined up over 10 sources. There was Little X, who we got over the phone from New York; Sol Guy, ex-manager of Figure IV; my Toronto school teacher friend, Mike Malcolm; Jemeni from Flow 93.5 FM; Phil Vassell, owner of *Word*; and others. From the jump, *Da Mix* was not just an hour-long vehicle to show new videos. I always believed that it should, on occasion, have some social relevance. Where this topic was concerned, I felt a certain social obligation because I used to give air time to some of the offending videos. It was important for

Much Master T

me to do this show because as, much as I used to play some of these videos, I was tired of the industry trends that created this artless, derogatory, promotional junk. I don't believe in censorship, but I felt the head honchos in the music industry owe it to the teens to explain what they're trying to do with these things, good or bad.

When I greeted India upstairs at Much, it was a comfortable meeting; she was cool. In this particular case I knew that I wouldn't have the intimacy of the interviewing process because a good deal of the emphasis for this show was on the panel. Her contribution to the show was just as valid and important though, because not only did she perform an amazing and most fitting live set, but she contributed to the overall discussion as well. I could tell she was really into it.

For the record, this show stands as one of my proudest moments as a vj because I was able to incorporate most of the elements of what my show was all about. We had youth involvement, music, and discussion. One of the most profound moments of the show came when I went to talk to the audience members. These young, teenaged girls began to speak openly and honestly about how, as young women with darker skin, they wondered why women who look like them so rarely appeared in the backdrop of most black music videos. We covered a wide range of topics and we came up with a few solutions to the issues surrounding the degradation of women in videos. This is what my show's always been about. I always wanted to give the urban community a meeting place where we could discuss what makes us tick.

In addition to the panel, we made use of some great clips from videos and past interviews with artists like Snoop and video directors like Hype Williams and Little X to get their take on the subject.

Celebrity Interviews

This show was info-heavy and educational. Plus, there was a live performance component, as India performed "Video" and "Brown Skin" to cap off the hour. It's funny because at first, India didn't want to perform "Brown Skin." Off camera she admitted that she was worried people might perceive it as a sexual song, which is why she gave a little disclaimer at the end. I let her make the decision. I wanted her to perform it because it was the first time Canada would get to hear her second single live. Still, she was worried about the viewers misinterpreting the lyrics. If anyone really took the time to listen to the words, I thought, they'd realize it was a song about love, not sex.

After the show, India told me that she wanted to come back and hang out on *Da Mix*. It was an exciting thought because I really wanted to do an in-depth, sitdown interview with her. Our show really had been something special. But somehow I felt it wasn't going to happen. By the time she would be ready to come back, I suspected, I would no longer be at Much.

Jay-Z and the Hard Knock Life Tour
(summer 1999)

Jay-Z is the self-proclaimed King of New York, King of Hip Hop and I'm certainly not going to dispute his claim to any of these titles. But that's not all that's impressive about this ent-rap-reneur. I hooked up with Jigga Man during what was arguably his biggest music industry coup

Much Master T

to date: The Hard Knock Life Tour in 1999. His Grammy-winning album, *Volume 2: Hard Knock Life*, was jumping off record store shelves and, at a time when recent rap concerts featuring Cash Money Millionaires had been marred by violence and concert venues were shying away from booking hip hop acts of any calibre, Jay-Z masterminded what might have been the most successful all-rap tour ever. The tour landed at the Air Canada Centre in Toronto after a stop in Montreal and it featured DMX, Method Man and Redman, and Jay-Z.

I arrived at the ACC after work. From the moment I entered the building, it was brimming with excitement. Thousands of adolescents were milling around the front entrance. I picked up my media credentials at the front gate and waited patiently for Jay-Z and DJ Clue to conduct some interviews with us. On first impression, Jay-Z was mild mannered and didactic. We exchanged pleasantries and then took a picture together. He was very observant of all of the activity going on around him: tour contest winners taking in the sights, and flash photographers looking for photo ops. For rap-starved Canadian audiences, this show was a big deal. Even bigger — and more wide-ranging — than I had expected. With more than 10,000 people in attendance, it was the first time I had seen such a large, ethnically diverse audience. There were an equal number of Asians, South Asians, blacks, and whites in the crowd, and that was something else. One of the beauties of hip hop is that, unlike most other music genres in the pop landscape, it crosses all ethnic, class, religious, and cultural boundaries. Also, this was the first time that Toronto had ever hosted a hip hop concert of this magnitude, so this was a night of firsts. Ten thousand hip hop heads in one venue. Unprecedented.

I interviewed DMX first and he's an interesting study in contrasts. DMX's albums all flaunt these really dark, morbid images that are

Celebrity Interviews

filled with blood and violent graphics. I asked him about this and the sense that I got (without him saying that much) is that hip hop is cathartic for him: it serves as a physical and emotional outlet. Point blank, he informed me that his music saved him, as he went through his life tribulations. He said hip hop provided a vehicle that he could use to channel all of these stray energies into. After wrapping up my DMX interview, I went outside and I wound up near the side of the stage in awe. DMX's show was nothing flashy. It didn't have much set design, but it wasn't supposed to. I was more geared towards the street audience. Meth and Red's set was bananas. They concluded their set suspended by wires hanging from the ceiling, zipping across the audience singing "How High." Jay-Z, sporting a Toronto Raptors jersey, went through his medley of hits with no pyrotechnics. It was all about his delivery; no dancing or shiny suits. He rapped through his catalogue of hits, finishing off the show with a wicked 45-minute set that most of the crowd knew the lyrics to. When the crowd started singing along to "Hard Knock Life," the show started to take on rock star proportions. I still get misty when I think about how the hip hop community came out to show love at a new venue that had just opened. I spoke to the promoter before the show and he was feeling the jitters. He admitted that he was taking a huge financial risk. But it was clear to everybody in the building that the gamble had paid off. Every corporate executive's kid was at this concert. The next generation was being weaned on hip hop and it had become clear that hip hop is a viable commodity that brings in a lot of dough. The show demonstrated that hip hop can be used as a unifying force that has the potential to surpass religion or money. The show also convinced the world that Toronto had arrived as one of the major world centres for urban culture.

Much Master T

After the concert I hung around to conduct a short but sweet Jay-Z interview in the dressing room. Jay-Z is a powerful man who demands respect. At one point during the interview his friends were whooping it up, making loud obnoxious noises. When he told them to clam up, they did. We rapped briefly about him being a black multi-millionaire. We also talked about his role as the leader of a new breed of young rappers that make a lot of money and are being called upon to represent the hip hop community in the political arena. We can judge Jay-Z for his lyrics and videos, and so we should, but it's what he does in the next quarter of the rap game that I'll be looking out for.

Snoop Doggy Dogg
(July 1997, February 2000, and August 2001)

Snoopy Doggy Dogg is, hands down, one of the most fun, over-the-top interview subjects I've ever hooked up with. The first time I interviewed him was at the Smokin Grooves Tour at Kingswood Music Theatre. This was his first time in Canada after the well-publicized murder of his bodyguard that he was alleged to be involved in. He was supposed to be under a lot of duress, but you sure couldn't tell during the interview. Snoop stands about six feet, four inches and he's lean, comical, like Scooby Doo. Snoop's Uncle Ray was there and he was telling me how he watches me on TV from L.A. Uncle Ray is the spitting image of Snoop: equally tall, lanky, and funny. A few weeks before I rapped with Snoop, reggae artist Super Cat was on my show. He travelled with a large entourage of followers and was telling me some tales of going to Snoop's house and smoking some Cheech and Chong–sized spliffs.

Celebrity Interviews

So I extracted this piece of info and brought it to the Snoop interview. When I told him about the Super Cat weed hook-up his eyes lit up. At this time, somebody was passing him a joint off camera and he was telling me how the joint was weak, not worthy, by his standards. While he was perusing the area for good weed, he was also ironing his pants and straightening out his perm for the show. All I kept thinking was, this guy is a card. Somebody should develop an animated series or a cartoon based on him because he's too funny for words.

The next time I hooked up with him was in Baton Rouge, Louisiana. He had signed a recording contract with No Limit and we were flown down to interview him along with Master P and C-Murder. Because my stage name sounds so much like Master P's, I knew I was gonna take some hits from the No Limit crew, and they didn't hesitate. From the moment I stepped into the interview area, the place was smoking like a chimney; a cloud greeted me as the crew took multiple hits from their bongs. Man, I was getting high off the second-hand blunt smoke. And they started making jokes about me being Master P's little cousin (for the record, we'd both carried this namesake for years; I'd become a Master in 1990).

This time around, it was clear to me that Snoop was reveling in Master P's business acumen, as he was anxious to start his own Dogg House Records label. Allegedly, he was being leaked for all of his funds through the murder trial. Death Row Records had also been ripping him off for years as far as royalty cheques were concerned. He was wearing the gold-plated No Limit emblem on his chest to signify his union with his new rap family, and for the first time I sensed that, underneath the cartoonish demeanour, there was a guy who was afraid for his life — thanks (at least in part) to the ubiquitous Suge Knight–fuelled Death Row threats.

Much Master T

The next time I hooked up with Snoop was in New York for *Tha Eastsidaz* release at the Interscope Records offices. Snoop is a modern-day Cheech and Chong, because man, the whole upper floor of this office complex stunk of weed. Even the label reps were coughing up blunt smoke. When Snoop finally came down for his interview session with me, he smoked some more weed and proceeded to waffle down some chicken in the process. Despite Snoop's drug antics, you could see a more mature, intuitive Snoop emerge over the years. The first release on his own Doggy Style Records was from a rap duo from his hood named Tha Eastsidaz. He explained to us how it was important for him to treat his artists fairly and have them receive all of their royalties. This was a tip he picked up after getting royally screwed at Death Row Records. Snoop was coming of age. By now he had the street smarts and business saavy to pull this label venture off.

Whenever I'd interview larger than life rap stars of Snoop's calibre I'd get them to do drops for my radio show, *Wall of Sound*. Over the years, DJ Dave Campbell and I had established a library of promo spots that couldn't be matched. The beauty of getting Snoop to contribute something to our library is that he's so unpredictable. He grabbed our recorder and started slurring his words, talking like a pimp. And the end of the drop, he said "was that gravy." Classic Snoop.

The last time I hooked up with Snoop, well, let's just say that he saved his best for last. I was at Reggae Sumfest 2001 in Jamaica, and during the press conference, there was this sudden torrential downpour. Who did I see when the rain cleared? Snoop Doggy Dogg. And just when I thought I had seen it all, I saw him and Kurupt wearing red, gold, and green Rasta tams, the ones with the fake locks attached. Here he was making a mockery of Rastas in Jamaica. When our eyes met, he looked over and waved at me. I looked behind me to see if it was me he was waving at, because you never know with Snoop. So we tried to nail an interview with him. We just walked and talked and conducted a short interview to recount some of his career highlights and lowlights. In true court jester fashion, Snoop told me that "everyt'ing is irie" (when in Jamaica speak like Jamaicans?), and he said how happy he was to get the chance to sport his dreadlocks (which were so obviously fake). I figured, this guy must be kidding. The funny part if this all is that, as Snoop left the interview area, the cameraman Tim told Paula, "Snoop has some nice locks; he grew them really fast, eh" and Paula and I burst out in laughter.

After our interview these two American journalists nudged their way up to us and requested Snoop's services for an interview. In true Snoop fashion, he launched into this fake Jamaican patois accent as a ploy to get away from the media. He told these other journalists to "leave me alone, mon." This guy is a riot.

Nelly Furtado
(March 1, 2001)

Before the Grammys, multi-platinum debut recordings, and *Rolling Stone* cover stories ("Whoa, Nelly"), Nelly Furtado was this cute, little hip hoppy chick who used to hang around the urban music scene performing at local showcases to get experience. In fact, many of the mainstream critics have no clue that she was discovered at this local

Much Master T

showcase for women in black music called Honey Jam, put on by Ebonnie Rowe and Phemphat, and was discovered by her manager, Chris Smith (a guy who had built a rep peddling reggae 45s, managing the Philosopher Kings, Prozzak, and Len Hammond among others).

Leading up to my interview with her I had heard such a buzz about Nelly that I knew Chris Smith and his management crew had done a good job marketing-wise. Nelly's name just seemed to roll off the tongues of pop music impresarios. I was at my doctor's office one day and I picked up a copy of *Elm Street* and it contained a story on Nelly. I made a point of reading the article to gain some insight into her musical style. I wanted to see just where she was coming from. I remember listening to her record with Paula a couple nights before the interview. We were both really impressed by her distinct vocal ability and the uniqueness of her entire sound. Immediately after Paula finished listening to the CD she said, "This kid's gonna be a star!" She has a knack for predicting impending stardom. For me it was interesting to hear the musical maturity and multi-cultural influences that Nelly was displaying. There was a jazz element, a hip hop influence, and a reggae fusion. This album could only have been made by somebody living in the most multicultural country on the planet. Her sound was all over the musical map, though it was pop at the end of the day. I was a bit disappointed in the fact that the media didn't focus on her hip hop roots. After all, she herself constantly pays homage to hip hop.

Leading up to the regular flow session with her, I had never heard about a live interview with her, so there was nothing to draw any inspiration from. I just came in with a fresh batch of questions. Her management team seemed to put the focus on the early stages of her career. They wanted to carefully control her image and map out

Celebrity Interviews

a fast-forward marketing strategy. Chris Smith carefully orchestrated the careers of his clients, and she had the Chris Smith imprint all over her.

When Furtado entered the studio I shook her hand. She admitted she was nervous because this was one of her first nationally televised interviews. Seeing the two of us standing side by side, it was a study in contrasts. Here was Nelly Furtado, the next young, big Canadian thing, and I was the older, seasoned vet on his way out. Just as I remembered from seeing her at those early performances, she was a giddy happy-go-lucky, free spirited girl with this hefty, childlike laugh. She was very down-to-earth. Her positive attitude toward life made me think that even though she may have come across some challenges in getting to where she was at this juncture, her attitude has probably helped her over the humps. On that particular day I was feeling far removed from Much. I was there physically, but not emotionally or mentally, but I was feeling okay with that. Nelly's carefree presence lightened my spirits, that's for sure.

You know you're getting up there in age when the pop stars of the day start telling you how they grew up watching you as a kid. But hooking up with Nelly before the career blowout that would see her performing for an Aretha Franklin tribute on VH-1, holding the mic at the 9/11 America tribute, touring with U2, and doing an AIDS benefit with Elton John, I didn't feel necessarily feel old: Nelly seemed mature beyond her years. When I asked her if she was getting nervous about the build-up, she smiled casually and said, "I am pretty grounded. I am really practical about stuff . . . the key thing is having patience. Patience is the greatest virtue in life anyway, you know. You let patience lead you through each day and you're all good; and when you remember to have fun, that's when you're laughing."

Much Master T

In America she's been painted as some overnight, out-of-nowhere success, but anyone who tuned in to my exclusive interview with her knows that she's no flash in the pan. She's been writing songs since she was 12, she used to dabble in penning hip hop rhymes with her Victoria homies, and she experimented with trip hop in a group called Nelstar. Although she's paid her dues, she still manages to hold on to her youthful exuberance. I sincerely hope she never takes notes from someone like Fiona Apple and loses that spunk, that youthful enthusiasm, becoming cynical about the industry like some of the other young female pop stars on the planet.

GOODBYE BLOCKO

Parting is such sweet sorrow. But how sweet it was to make my exit with Lauryn Hill taking me out with a bang after 17 long years at Much. It was a divine order that enabled Petal and I to pull off one of the biggest coups in the history of departures at Much. Many people in the industry came out to show their support and all the viewers that made the trip down to see me off in grand style were graciously entertained by some of the hottest acts in the Canadian music scene, stretching from the past into the present. It was an intense month leading up to the big day, but I was more than ready for it all to go down. Unfortunately, I wasn't ready for everything that would happen in those few days.

So how did I land the world's most popular borderline recluse

Celebrity Interviews

and star performer for a gig at my Much "Goodbye Blocko"? By pushing envelopes rather than licking 'em. Keidi-Ann from the BSA (Black Students Alliance) had already been trying to bring Lauryn to Toronto. I had a good connection with Jackie from Kings of Kings Promotions, who would have backed the show. For my 10th anniversary blocko, Jackie had helped me to land the services of Mr. Vegas to perform, so I asked her if there was any hot performer she might have available, and she said she'd think about it. Then Jackie suggested that they were already trying to bring Hill in on August 25th. Keidi-Ann informed me that she would not be booking this gig through Lauryn's label, Columbia/Sony but that this would be a below-the-radar, one-to-one deal. At this stage in her career, Lauryn was going through some issues with her label and personal issues with Rohan, so she was only doing shows that she wanted to do and was not interested in performing old songs. Keidi-Ann suggested that I e-mail Lauryn and let her know that I'd like to use her services for my blocko. So I wrote out my spiel: I'm leaving Much after 12 years, I hope you remember me from my interviews with you in the past, yada, yada. I sent this e-mail out but it had been a while already and I still hadn't heard anything back from her. I began to think that she probably couldn't accommodate my requests. The Canadian labels here were always trying to get their largely American rosters to come up and perform, and this created a weird predicament. If the American artists weren't going to come up with the support of their labels, it would never happen (except for a promotional tour).

I re-focused my energy on putting together a good, strong Canadian roster. I wanted to get an even mix of artists that represented my 12 years on air. I needed to get the acts that had laid the foundation for this urban music scene, and that meant landing

Much Master T

Maestro and Michie Mee. As far as the new school acts I assembled, some of them were bubbling performers about to blow up. Acts like Jelleestone, Lenn Hammond, K-os, Baby Blue, Choclair, and Mr. Mims. It was important to land these up-and-comers because I always viewed my show as a showcase for new, emerging talents. Now all I needed was a special guest to seal it off.

This week, I had to go to Jamaica to cover Reggae Sumfest 2001. And I had an acute case of Lauryn on the brain, and I was trying to contact Keidi-Ann to get a progress report. I came back to the hotel after Dancehall night, which was nuts. For starters, the after-hours performances were in motion, which meant that I was literally conducting interviews all night. As well, Sumfest is a long, gruelling festival to shoot. Beenie Man and Bounty Killer were feuding and bottles were being thrown at the stage. We even heard a bunch of gunshots go off; the Much crew and I panicked and could easily have been trampled. After literally fleeing for our lives, I went back to the hotel and caught some sleep.

The next morning at 8 a.m. I received a phone call from Margaret at Much, who's responsible for booking flights and schedules. She told me that I should get in touch with David Kines because Lauryn had called him. She implored me to call Lauryn as well, and she advised me to sound a little bit more excited about the prospects of her coming to Toronto than David. Apparently Lauryn told Kines that she'd be interested in performing at my send-off and he was like, "okay, cool" in a melodramatic fashion. And Lauryn responded by replying, "okay, cool" and that's it. Kines seemed ungrateful in her eyes. So I called Kines and he then called the record label and the label said they'd take care of the flights. Then I called Lauryn personally. She was in Jamaica and actually wanted to

Celebrity Interviews

perform at Sumfest, but Sumfest organizers were nervous because she was only performing new tracks and they weren't sure how receptive the audience would be to this.

So I got into this nice conversation with her while she was staying at the Ritz Carlton. She told me how things had changed for her and how she viewed the record label in a different light and wanted to take control of her own affairs, right down to eliminating a staff of 70 people when she toured. She was feeling the strain of having to carry all of these paycheques. Plus, she was beginning to understand what the industry was really about. It was all about creative control and Sony wasn't interested in these new songs because she didn't have an album out. I asked her about the rumours concerning her and Rohan not being together and she said that they had things to work out as a couple. I asked her about the publicity surrounding her performance at the 2001 Essence Awards where she broke down and cried. She explained that the songs she was performing were very personal and made her emotional.

The following night I just happened to run into Lauryn's hubbie, Rohan Marley, at Sumfest. He asked me if I had called Lauryn yet. She had now left the festival so I called her New Jersey number and sent her an e-mail. I couldn't get in touch with her and we were only two weeks away from my send-off. And that wasn't all that concerned me. The Kings of Kings event fell through because the BSA and Keidi-Ann couldn't lock in a venue to support Lauryn's specs (she requested to perform outdoors).

When I was dealing with the negotiations I was also starting to realize that her potential performance at my Blocko would definitely not be about paying homage to me. It was this unwritten deal where she would do her songs and I could use her appearance

Much Master T

as a promotional tool for my going-away party. So I called Kines and he said that we'd have to be responsible for putting them up in the hotel and arranging their flights because the local label affiliate would have no involvement. So the pressure was on.

Much was wondering whether this event was still happening. Sony and Columbia refused to pay for the flights (or anything else for that matter). Much finally agreed to pick up the tab. However, Lauryn was incognito and the cell number I tried to reach her at had a voicemail message that said "the cellular customer is not available." Finally, one night I called this same cell number and I heard reggae music playing in the background. It was Rohan jamming and he proceeded to put Lauryn on the phone. "Let's make your event happen," she said. She put me in touch with Addis, who's Ziggy Marley's road manager, and that made my life much easier because he had Lauryn's direct number (the numbers I had were for the Marley Museum in Kingston and Strawberry Hill, Chris Blackwell's property). So all of this last-minute jockeying and preparation was taking place on the Wednesday and the event was scheduled for that Saturday. In an 11th hour move, Sony Canada agreed to take care of the hotel fees, which I thought was a good thing, at first. But Lauryn didn't want the label to be involved and she kept stressing that she was under no obligation to do anything with them.

Without the label's involvement, I had to deal with finding accommodations, meeting sound and stage requirements, getting her a wardrobe and a stylist. And oh, did I forget to mention that she was six months pregnant and that her two kids (Selah and Zion) were coming? She needed five plane tickets because her personal assistant was booked on a flight as well. So we hit Thursday, two days before my event, and we found out that they needed to have travel visas

Celebrity Interviews

from Jamaica to be able to get her people to Toronto. Neither Sony nor Much had ever dealt with a visa from Jamaica, so I had to make some quick decisions. I called a friend of mine, Nick Cumberbatch from Winnipeg, who had imported many reggae artists to his hometown. Nick was a reputable promoter and he knew somebody at the Jamaican embassy. So after some minor haggling, we ended up getting two visas on the Thursday. But we weren't clear yet. The day of the show, we found out that the one girl who was supposed to come up had been deported from Canada six months ago.

More drama. Then Hill's people decided that they wanted to bring up a guitar tech at the last minute. I explained to Hill's camp how hard it would be to go back to the Jamaican consulate and request another visa for a band member (considering that someone in her camp had just been deported days earlier). Anyway, we got the guitar tech into the country and, one year later, I still have no idea how. So Hill and her entourage flew in on the Friday and they were on the only flight out of Jamaica at 2:45 p.m. We still had to work out with Addis who was going to get paid. We came up with a scheme whereby the road manager would receive compensation, and so would the engineer (who was the same guy Shaggy uses). We had to be as detail-oriented as they come. Even though Lauryn was performing an acoustic set, she had specific sound requirements so we had to hire two guitar techs on our end.

As Lauryn flew in to Toronto, Petal and I headed to the airport. The flight arrived, and one hour later we spotted them. I had just picked my son up from daycare and he was really exhausted, so I had to be wary of that. When Lauryn stepped off the plane, she also had a handle of both of her kids, Selah in one hand, Zion in the other, and walked out exuding a certain confidence.

Much Master T

Mega superstars have mega specs when it comes to travel arrangements, and this provided a whole new set of challenges when it came to providing the vehicle that would take her to the station. Hill did not want to be carted around in a limo. Because we couldn't get an SUV in time, we had to rent this stretch limo. I remember her going into the car and pulling Petal aside and explaining that she didn't want to use this car on the day of the show. So there was this motor brigade of two black vans and this limo heading to MuchMusic and Petal and I were looking at one another and thinking "this is surreal."

After all of these very adult negotiation sessions, Zion, Selah, and Kalif were playing together in the Much set-up room. Petal offered Lauryn some tea and she accepted, as long as it came with some honey. As far as styling issues went, I hooked up with Daymon from Lounge Clothing at Queen and John and he put me in touch with a stylist named Donovan. They presented all of these outfits to Lauryn. At the end of the day she didn't wear any of the gear that they presented; she wound up wearing one of the outfits she had brought from New Jersey.

By Friday night, I was floating on air, knowing that Lauryn Hill had arrived in the country with no problems and that she'd help send me off. Kalif was overtired. He usually goes to bed by 9 p.m. and it was 10 p.m. Hill came down for the soundcheck and a crowd came from out of nowhere. Nobody believed it was really her (again, she had only performed twice in the last year, anywhere). American journalists had no clue she was here. I don't think there was anything more incredible than the music cutting through the night air for soundcheck while people were gathering at the parking lot. Most of the Much employees were just taking it in as well. She was

Celebrity Interviews

using the people gathered as a sounding board for her soundcheck. When she finished the soundcheck an hour later, I took her back upstairs and waved goodbye. It was close to midnight and Kalif needed to get to bed, so I had to expedite some of our treatments of the sound problems and mosey on home. The problem was, we didn't have an in-ear monitor, so we had to re-do the soundcheck with Lauryn in the morning. This could have posed a huge problem because we were slated to have six other soundchecks (for the Canadian acts) to go on that same morning.

On this grandiose day, Petal, my trusty assistant, was running around trying to ensure that the day went off without a hitch. Rascalz' Red 1 presented me with an engraved "fist of power" trophy on behalf of Figure IV Entertainment, Canada's premiere hip hop company. Shaquille O'Neal, Redman, Lenny Kravitz, Master P and son Lil' Romeo, Vince Carter, and others sent best wishes via video. As we were getting close to the performance, Lauryn arrived on the stage a little earlier than expected, sporting a vintage Bob Marley T-shirt, which was good because she got to talk to and prep the crowd for what she was about to pull off. She kept insisting that the crowd should peep the lyrics and ignore the glam and pomp surrounding her performance. She explained how she finds grace through her music. She was performing all new material and she was insistent that we pay attention to the subject matter as opposed to her outfit or make-up. It wasn't until she sang the first song that people realized this was no Lauryn Hill look-alike clone. Kines wanted us to explain to the audience that Lauryn is here, and I urged him to relax.

There was this nervousness and trepidation because we had to stick to the live format and she had to stick to her time allotted. She was to start at 4:15 p.m. and her set to end at 5:00 p.m., to coincide

Much Master T

with *Da Mix* signing off. I had asked her in advance how long her set would be and she explained that she didn't really have a set. And I was like, okay. I informed her that I'd like to take the time, just minutes before show's end, to say my goodbyes.

As it turned out, Hill did end up going over time and Kines started breathing down my neck to wrap it up. I made a rash decision because, in my heart, I knew it was important for her to continue on. Plus, I wanted people to be captured by the performance and not be put off because the lyrical content was way too much to digest. So Hill ended up performing a 45-minute set, which I don't think anyone expected: it was like a mini-concert with an acoustic guitar. And this was the most intense lyrical flow anyone had heard in a long time. As I was taking it in, I talked to Addis who stood there and said, in his Ethiopian accent, "She has such a beautiful voice, so unique." It was an emotional moment for me and the lyrics hit home. My fans took it all in, and walked away with those lyrics still drifting through the air.

I was as happy as a rat on Spadina. In the end, it was emotional for me for too many reasons. This was the first time my whole family had come to the Much building together. I asked my foundation to come to the stage to let everyone see and appreciate all the important people in my life that have stood by me and kept me strong. My wife and son, Basil and his wife Yvette, my sister-in-law Donna, my brother-in-law Kurt with my nephew Tarin, Dave Campbell, and Dawn Craig (a close family friend) all lined the stage to close the blocko out. My mom and Aunt Sis were in a private area in the audience and the cameras showed cutaways of their proud, smiling faces. It was the first time folks saw my wife Paula on-air. And to see some of my Much comrades crying, like Laurie, who had

Celebrity Interviews

been doing my make-up since my *X-Tendamix* days, put me over the edge. Just to see how she was affected by this sudden change meant the world to me. That's camaraderie you can't buy.

Lauryn's performance echoed the feeling out in the audience that day and from all accounts it was a melancholy vibe. Sadly though, that air of melancholy would take on a new meaning when we ran into Little X that night as we arrived for a little soiree in my honour. He told us, to our disbelief, that Aaliyah's short life had come to a tragic end in a plane crash that very afternoon. I just remember Little X getting frantic, wondering if Hype Williams (his mentor) was also on that same plane that went down in Bermuda. I kept having these Aaliyah flashbacks. I thought about the first time I interviewed her, about her quiet charm and understated aura. Very few knew that Aaliyah was slated to be a guest host on *RapCity* and *Da Mix* for the week after my departure. It was a poignant and shocking evening for me, and that day will forever be etched in my memory.

The goodbye blocko was emotional for me in a different way because, in my heart of hearts, I had started to do the work, the healing work that was necessary for me as a man to achieve some inner growth. It marked all the accomplishments that I (and so many others in my personal foundation) had achieved, and I was ready to dream a new dream and take a leap into a brand new chapter in my life.

Much Master T

LIFE AFTER MUCH

When the word got out on August 8, 2001, that I would no longer be a Much employee and that *Da Mix* might ultimately be getting sent to meet *The Beachcombers, King of Kensington,* and *Drop The Beat* in the great studio in the sky, it was not the best day. I hosted my last show at MuchMusic on August 25 and at that point it was time to dot the I and cross the T, no pun intended. My career at Much was finito. Kaput. Awkwardly outta here. I wanted to bow out gracefully, not be the VJ that people thought got the heave ho. To use a soccer analogy, I wanted to be like Pele. When Pele retired, his general attitude was "Okay, I'm a great soccer player. I'm done." In a similar vein, I believe I left MuchMusic at the top of my game. I had a great run as a VJ and now I'm finished.

People ask me why I left, and whether it was amidst scandal and controversy like Ziggy Lorenc, but it wasn't that. I just knew I had to leave. It was the way it happened that soured me. I wanted to move from downstairs to upstairs, and I wanted to be an executive there, that's it. That didn't happen for a plethora of reasons and I'm not sure if there was a glass ceiling or what. I've never looked at my future more than I have now. This pop culture fame stuff is great, but you know what, you can't live on that. You can't live on people patting you on the back. For me, it's always been about pushing urban music and culture in Canada. I think, with my show, that's the one thing that people were always able to rely on. Being one of a handful of black hosts on Canadian TV meant that the African diaspora and general mainstream music community could look towards me for their reggae, rap, and R&B fix and say "there's some-

Celebrity Interviews

body who's proud of their heritage, who wants to celebrate it and make it universal." This formula has worked. Urban music is definitely not just a black phenomenon; youth of all persuasions are buying it by the bushel and I think I've been a big part of that. After working anywhere for 17 years I would like to think that I had a stake in something.

It's funny, but in the rarefied halls of Much you barely notice the passing of time. Like any job, if you enjoy it, time is nothing, or maybe it's absolute. That's what happened to me. I didn't feel like I'd been there since 1984. In fact, the more time I spent on camera actually helped me settle into a comfortable groove when I was handling the biggies — the world exclusive interviews with stars like Madonna and Janet Jackson. I think my maturity and professional growth helped develop my interviewing style. It takes a lot of time to find yourself and to discover the different ways you can ask questions and make people feel comfortable in front of a camera. A decade or two may not seem like much in the real world, but in the fickle world of music TV, it's an eternity. I've seen a whole host of personalities come and go, from Erica Ehm to Christopher Ward to Steve Anthony. Others have used the revolving door at Much to reach dizzying new heights. Ehm, for example, went on to a successful career as a songwriter, broadcaster, and public speaker. Rebecca Rankin, a videographer-turned-VJ, joined U.S. adult music station VH-1 in New York as a globe-trotting reporter. And former Much teen idol J.D. Roberts is a Washington correspondent for CBS News. Laurie Brown, former host of *The New Music*, is now a CBC arts and culture reporter. Of that class of VJs from the 1980s, I was the last guy standing.

I had to ask myself some tough questions. Was I being politely forced out because I was 40 years old in an industry populated by

Much Master T

twentysomethings? Or was this a payroll procedure, and had I reached my salary cap? Without getting into any details, for the moolah I was making at the time, Much could have hired two new VJs and two junior producers. When I left, they did.

What really went down came about when I went in to negotiate my contract in August 2000. It was suggested that I stay on only until December 2000 and I was shocked. For the first time I knew that I really needed a lawyer, and after some negotiations took place it was decided that I would stay on until August 2001. During the back-and-forth discussions the reality of what was happening began to sink in, and I tried on several occasions to renegotiate. I suggested coming off the air and creating an urban department (one that was well overdue), including *Da Mix*, *Rap City*, and *Much Vibe*, but to no avail.

The writing was on the wall, because on the exact day of my 40th birthday in May 2001, I received a few omens. I'm a man who's built a career around being organized and disciplined and the fact that I lost both my cellphone and the keys to my truck on this birthday made me think further about taking stock of my life. Another omen occurred when I attended a goodbye party for former MuchMusic head honcho Denise Donlon. She said something along the lines of, "T, this is your time now. The doors are open wide for you." At the time I didn't understand why. I replied, "Yeah, yeah, whatever Denise. You just wanted me out." Donlon was incensed over this comment when it ran in the *National Post*. What really happened, and got misinterpreted in the *Post*, is that my comments were not said aloud, but were private comments I was saying to myself in my head. It took me a few months to fully respect and understand where Denise was coming from.

My last three months at Much were an emotional journey, and

Celebrity Interviews

these weird ideas started creeping into my head. Maybe I was feeling old, but at the same time, I was feeling more youthful because of the zest and spontaneity I had gained. I've since gone through the post-MuchMusic withdrawal symptoms and sometimes I miss the energy of the job, but at the same time, there's a new sense of freedom. A *Kitchener-Waterloo Record* reporter who interviewed me just following my retirement announcement wrote that I looked "tired — almost like a zombie going through the motions." They say two of the biggest stresses you can go through in life are changing careers and moving. I was going through both and Paula had grown a little tired of Master T. I needed some mental cleansing and in June 2001, the 5th Annual St. Kitts Music Festival would act as that sanctuary.

St. Kitts is one of our favourite islands in the West Indies and normally Paula would have joined me on a wonderful excursion like this. But not even the ocean, sun, and fancy drinks could convince her to come. After covering the four-day festival that featured K-Ci and Jo Jo, Chante Moore, Sean Paul, and Burning Spear, I was emotionally and physically pooped. The day before I was to depart back to Toronto I decided to take a little time for myself. My room at the Jack Tar Village had a wonderful view of a golf course, a lake, a wedding chapel on the edge of the water. The only ceremony I performed that day, however, was a pledge to my journal, dated June 24: "The picture becomes clearer and clearer, I will survive." When I arrived back in Toronto, a major part of my stress had been lifted.

When I saw the first set of hyper-slick ads running on MuchMusic for MuchVibe, a music-video channel for fans of hip hop and R&B that promised "dope beats," "soul sisters," and "fresh MCs," I wondered why I wasn't involved in the new channel, despite my major role in pushing MuchMusic to play more black music

Much Master T

over the years and my stature in the Toronto urban music community. The 24-hour urban music station came to be as a result of the blood, sweat, and tears of myself and folks like Michael Williams, Michelle Geister, Siobhan, Sandra, and Petal.

I mentally started to distance myself from all things MuchMusic and began to take solace in my other non–Much-related accomplishments. For example, throughout my years at MuchMusic I received numerous awards for various deeds well done, but in retrospect, very few of them were as gratifying and prestigious as my Mohawk College 1998 Alumni of Distinction Award for Applied Arts. The moment I received this award ranks right up there with my top interviews, especially for my mother who saw me through my first years at college. Man, when my picture was put up on a Wall of Distinction at the Fennell Campus at Mohawk College on January 22, 1999, it was an emotional event. I think Mohawk College's media studies programs should take some of the credit for my success. It must be one of the best programs in the land. Look at what it helped me accomplish! I will always encourage kids considering post-secondary education to choose Mohawk. Why? Because I did.

What does the future hold for me? Unfortunately most VJs aren't considered "real" journalists, so we tend to get pigeonholed into oblivion. The vast majority of VJs don't reappear on mainstream Canadian TV working for competing networks — that's for sure. As far as a black TV dude with dreadlocks being welcomed to work on networks that are doing a questionable job representing shifting demographic realities? Let's just say that I'm not sitting by the phone waiting for any of the networks to pony up any job opportunities soon. The truth of the matter is that in TV Land, opportunities for people of colour are limited. That's why I'm looking to produce my

Celebrity Interviews

own shows for Canadian airwaves.

Also, a lot of people don't know this, but just weeks before the official date I left Much, I was approached by various record labels to come on board as their prized new recruit. BMG approached me first through urban music impresario Ivan Berry (think Dream Warriors, Michie Mee, and Organized Rhyme's Tom Green). Some of the labels openly wondered what kind of juice I'd have after leaving Much, but Berry, who is way more knowledgeable than most major label reps, argued that I still have plenty of value. He'd tell me that he had to remind them that I was the *Dance Mix* pitchdude — a guy who has diamond sales certifications to his credit. And speaking of *Dance Mix*, I also got into initial talks with my old compadre Dennis Garces from Quality Records, who is now at Sony Records. After some discussions, however, I didn't think Sony would be the right fit at this juncture.

My first and most important meeting, as it turned out, was with Deane Cameron of EMI. Craig "Big C" Mannix, the urban product manager at Virgin Records, had mentioned that EMI was interested in me. At our first meeting, I was thinking that EMI was looking like a pretty good place to be, because Deane more than understood my skill set in terms of TV and radio production, as well as my ability to brand compilations. My negotiations with Virgin/EMI started in October 2001 and the deal was finally inked in February 2002. It was an emotional moment because there was an almost four-month period of time that had passed and I was anxious to get things going. When I finally inked the deal, however, it was like I had started a new chapter in my life. I signed a six-album deal with Virgin to put out Master T–branded compilations of all black music genres from reggae and hip hop to R&B, soca, and neo-classic soul. According to

Much Master T

the contract, I have a development deal where I can include three artists of my choice on each compilation. When I started out in the music industry, my goal was to expand the urban music scene in Canada and to help nurture local artists, and my recording contract with Virgin allows me to do just that. Also, over the next few years I plan to expose the nation to Master T–branded DVD through my production company, Fullstedd, and my boutique record label, Serro Records (a nickname my father was given as a child). All this will give me the opportunity to develop some of the Canadian product I've believed in over the years.

Landing gigs when you're in your 40s, after having worked for one company for 17 years, ain't easy. For example, when you work with a major label versus a media house, the pacing is slightly different. When I left Much I was working two VJ shifts a week, producing and hosting *Da Mix* and *Rap City*, as well as programming a radio show. My hands were completely full. In November 2001, when Warner Music Canada contacted me to interview Quincy Jones at an Indigo bookstore, I was actually nervous. After all, it was my first post-Much gig. I had interviewed Quincy before, heck, I'd interviewed almost every major world artist before, but I felt just as jittery as when I conducted my first interview at MuchMusic.

After having received so much public exposure, I almost became a recluse. I felt paranoid, feeling that folks were watching my every move. The post-MuchMusic withdrawal symptoms kicked in big-time whenever I'd go to the grocery store or out to eat. At the Quincy Jones interview, over 200 people showed up and Indigo head Heather Reisman was there. I felt like I was back in my element when the crowd started roaring. A similar epiphany occurred just days after I signed the deal with Virgin Records. After I drove up to their head

Celebrity Interviews

offices in Mississauga and did the photo shoot, I woke up the next day in a cold sweat, realizing that I didn't have to go in to Much. It was a sense of freedom that I hadn't tasted for over 20 years. I finally realized that I hadn't taken a break since entering college. It was starting to feel good!

I'd say that the next major career move I'm looking to make is to produce a two-hour syndicated radio show in a countdown format. Just before I left Much, I ended my run on Z103.5 as host of *Master T's Wall of Sound* because it was getting to be too taxing producing the show in addition to my regular Much duties. Ironically, after an article was published about me in the *National Post*, my ex-Much VJ buddy Steve Anthony called me to tell me that Standard Broadcasting CEO Gary Slaight wanted to rap to me about some possibilities at Flow 93.5 FM, but that's still in the works.

At the end of the day, my whole vibe, ambition, and driving force was to capture the energy of the streets and transfer that to the Much airwaves. That's why I would always send a "big up" to our vast Canadian audience. I'd always try to embrace people from all walks of life as my extended "urban family" — that's why I continually addressed my audience as "nation" or referred to my viewership as "cool cats." Without tooting the horn of myself and my loyal audience, it's a good feeling to know that I and the *Da Mix* diehards have been a part of the success of Much and the growth of urban music in Canada. I'll be aggressively trying to plug all of my friends, well-wishers, and fans, into my Web venture at tmediamachine.com, a one-stop online urban portal that will give Canadian teens, adults, and everyone in between an opportunity to develop their entrepreneurial skills or just chat about the state of urban music in Canada on our message boards. I guess there is life after Much, after all.

About the Authors

Dalton Higgins is one of the most provocative observers of pop culture in Canada. His works have appeared in *The Source* magazine, *Vibe* magazine, *Amazon.com*, *Saturday Night*, *Now*, the *Toronto Star* and on CBC Radio One. The former editor of *Word* magazine was also a pundit on CTV's (Talk TV) *The Chatroom*.

Tony Young (a.k.a. Master T) will be releasing a series of Master T-branded CD compilations on EMI/Virgin Records and DVDs from his own Fullstedd production company. His music label Serro Records will be developing Canadian talent and his Web site *Tmediamachine.com* is a one-stop urban Web portal that will help Canadian entrepreneurs market their wares.